COINCIDENTLY-QUANTUM

NO MATHS, NO GRAPHS, NO DEAD CATS!

Author: John Blackall.

TABLE OF CONTENTS

INTRODUCTION

Science is a wonderful thing. I love it and through this book so will you. It is written so that if you, like many others, find science is frustratingly complicated with graphs, maths and diagrams you will delighted by this refreshing and easy to read summation of the way in which quantum theory has changed forever the way in which we view reality.

You will look anew at the idea of other dimensions, the Big Bang, the human brain and time itself. You may not understand how something can be in two places at once, move faster than light, nor time can go backwards, but you will understand that nobody knows the answer to these questions anyway. Above all you will know what an incredible chapter of history you are living through.

It is very clear that the world of physics has come to a crossroads not only between the two competing theories of relativity and quantum theory but also the possibility that we need a completely new direction to look in if we are to make any sense of the new and amazing discoveries, which are now almost daily. In order to put these pieces together we have to start thinking outside the box, mix philosophy with research and experience, and be prepared to think the absurd.

However, before we can start to make sense of the very real possibility of other dimensions, and start to share in the incredible and awesome developments taking place we need to familiarise ourselves with what is known as the Standard Model, quantum theory and some of the fundamentals of physics. To achieve this we will assume that coincidence is involved in some way we have yet to discover. That is an exciting idea and will allow us to stay in touch with the latest developments. It may be unpopular with those who treat science as static and unchanging but that matters not as one of the things we know is that every scientist worth his salt from John Wheeler to Richard Feynman will tell you again and again that nobody knows the answer to the most important questions anyway.

So sit back, take your time and check it all out by visiting any of the sites referred to and placed at the back of the book. I have numbered each issue or theory you may want to question, explore or verify. You will find the reference for each number in the relevant chapter in the Bibliography.

John Blackall BEd., DIPed., Scientific Theorist.

CHAPTER ONE

COINCIDENCE

I have never believed in ghosts. I am prepared
to but as nothing in the remotest has ever happened
to me which suggests there are entities and
experiences out there which are supernatural I have
always remained a disbeliever. However, I have
heard stories off reputable people, people who
clearly believe something has happened which
defies a rational explanation and that has served to
help me be tolerant of others beliefs. Consequently,
I have never denied the possibility that there are
things beyond my own experience, but until
something happens that convinces me otherwise I
must remain a disbeliever about ghosts, UFO's, the
Loch Ness Monster and many other unexplained
incidents. This belief no longer applies to
coincidences. Other experiences have convinced me
that something else is at work and that is what this
book is about.

Although I have always kept an open mind this
is not to say my views have never been put to the
test and that I have managed to get this far without
being in awe of the miracle of life and the wonder of
existence. I believe that I am no different than the
rest of mankind and just like everybody else there
are a number of occurrences in my live which have
left a mark on my thinking and have made me

realize that we still have a lot to learn – if indeed it is possible to learn much more about the underlying reality than we know already.

The first such incident is from when I was a child and I suddenly had this amazing awareness of my own presence. I was about four or five years of age and wanted to shout out to the other people in the room that I was here, not because they were not aware of me but because I had suddenly become aware of myself and my surroundings. This thought among all others has stayed with me all my life – I think it is called growing up!

The second experience is very similar. I was a little older, about thirteen or fourteen. I was in my grandmothers drinking tea and sitting on her sofa alongside my brother. Then, just for a moment, I could see myself from behind. The whole room and the people in it clicked into view like a picture and I felt that I would one day see this picture again. It was not just a visual experience – it was packed out with feeling. To this day I can still relate with awe and fascination to that moment in time.

The third experience involved a very distressing time in my life and is far too difficult a tale to relate in full. Suffice it to say it was somebody else having a similar experience, somebody very close to me. He told me of a moment in the near future and explained graphically that he could see himself in a box with people staring down at him. My memory of the funeral and the unfolding barrenness and coldness of the cemetery chapel had me relive those premonitions only days later.

These events have undoubtedly been a subliminal factor in accounting for my current beliefs, although they fade into insignificance when contrasted with the experiences of many other people I speak to. The point I am trying to make is that you would be wrong if you thought it accounted for what you may think is an obsession with theorising about coincidences. That would be very wrong. Just like everybody else I have ever known I had no reason whatsoever to question the common-sense view until I started to read about developments in science and in particular quantum theory. Its incredible claims and their implications opened up my mind and set me to question an assumption we all take for granted every day of our lives, that a coincidence is simply a coincidence. It reinvigorated the belief, as Shakespeare wrote in Hamlet, 'there are more things in heaven and earth, Horatio, than are dreamt of in your philosophy'. And so it is an ironical coincidence, or a gifted in-sight that my readings on quantum theory and a series of minor events set me on this journey, not a predisposition to reactionary and hysterical views

If you like me are in awe at the sheer wonder of it all then you may find this article of interest. There are a massive amount of strange and incomprehensible coincidences and a huge amount of books narrating them. However, I want to start with the ordinary and the familiar. And with that in mind I will proceed to some of the coincidences, we are all have experience of, which have helped concentrate my mind on the issues.

My interest started sometime in my forties on a regular visit we made to see my stepdaughter in a residential home for epileptics in the middle of Cheshire. On route there was a bridge which allowed for only one vehicle at a time. Neither vehicle can see what is on the other side of bridge and must proceed very cautiously – often at some point sounding the horn. Over time, despite this being a remote area of the country, we could not fail to notice the number of times we encountered an oncoming car. Amazingly it was less likely to happen if we anticipated it happening.

After a while we became aware that it was a little like looking or not looking. At first we would joke about it calling it 'quantum', but after a while it became a much more useful label than the concept of coincidence by making us aware of how incomplete and unsatisfactory an idea coincidence could be in relation to some events.

I am not assuming here that the 'Bridge Incident' is anything more than the result of my exasperation at not having things my own way. However, coincidence is a term we all use to embrace a wide range of happenings and there are many which leave us in amazement at their occurrence. In order to avoid the task of making too fine a discrimination I intend to refer only to what I term subjectively as the ordinary and the extraordinary.

The following is one of my many examples of the ordinary. I am writing this while on holiday and on the way back to our holiday home today I made a few notes myself. We enter an estate some 15 mile along a narrow two lane highway. The joy of riding along this road is that there are always very few cars on it and most of them will overtake you if you stick to the speed limit. The lack of traffic and the propensity for other drivers to overtake means you can meander at leisure along about 30 mile of beautiful coastal road. And so we did. At the point where we needed to turn two other cars, from two directions, one behind us and one coming towards us also decided this was their turning. The estate we were entering we later found to have no more than about 10 families on it. So adding the families on the estate and the traffic on the road does not seem commensurate with three cars turning into the same entrance at the same time.

We can of course spend some time on this looking for a rational answer to this motoring incident and I have no doubt we would find one. We always do but we still wonder at it.

We would also find an answer for another seemingly innocuous event that day. On the way home we stopped at an ice cream parlour next to a small shop. At the time we were the only vehicle in the car park but on leaving found the car park almost full. This later example is either a coincidence or our arrival there coincided with the optimum time for buying an ice cream - either answer would be a normal conclusion.

Such coincidences are very common to us all and the term coincidence perfectly sums them up for us without any fuss. The following day to day ordinary example is another good example of this. I swim regularly and through my activities I have become aware of a pattern which has given me the germ of an idea. That is what this article is about- a strange and wonderful idea! For a long time I found that I could almost predict that when I returned to my locker to change into my clothes there would be somebody else having to use the next locker. Not very remarkable you may say. However, considering that there were some 400 lockers in the room and on the majority of occasions only two people, not a 100 or more, were getting changed then chance was behaving with the predictability everybody who went to the sports centre was familiar with. In fact I often mentioned it to whoever was alongside me, usually with a knowing glance, a half grin or sometimes a passing comment such as, 'it works every time' or 'it never fails'. Nobody ever asked what I was talking about. It was always accepted that something was happening that common sense did not allow for.

We all recognise that what we see as a coincidence can be a very personal and subjective experience. Each of the examples I have given can be seen in this way and each of them can be explained away by applying a little logic. For example, we were driving through Wales some years ago. We were on a very straight road which dipped in the middle. This meant that you could see ahead for about two miles. In the distance was a car and heading towards this car was myself and a lone cyclist. I remarked to my wife that I had this feeling that there was a high probability that we would all pass each other at the same spot – and sure enough we did.

This is a good example of how the mind becomes a key player in making something out of nothing. The first thing to note about my example is that this is only a coincidence in terms of what I predicted may happen and what actually happened. It stems from a drivers concern when trying to anticipate any problems ahead when driving. The usual and more 'sensible' explanation is that one or both drivers are aware of the cyclist and subconsciously, out of their concern, make adjustments to their driving which account for the event occurring in this way – which ironically is the very opposite of what is wanted. Without having a feeling in this situation that what we predicted was about to occur then we would of course have no awareness of a coincidence occurring. In other words the thought itself has a role to play in what we call coincidence.

In all the examples I have given so far the human mind can be seen as a key player. This is true even for the story about the lockers in the changing room. It had reached a point that each time I returned to my locker I was already anticipating the likelihood that somebody else would be there sharing my space. Mentally I may have recorded that but not the number of times it did not happen. In other words, a form of paranoia.

All very ordinary stuff we can explain and given a choice we could have just as easily referred to these examples by saying quite honestly we do not understand but there will be an explanation.

However, there are other times when we find that there is no easy answer, and the word coincidence, is all we have. At such times we have no other word. A 'coincidence' is all we have.

So let's get more serious and have a look at some examples. One such happened to me some thirty years ago, but I will never forget it as it is one you rush home and tell the wife about, stating with the phrase, "You won't believe this". It related to two other people and I was possibly the only witness to it. I was acting as a taxi driver. I sat in line at the station for some 30 minutes awaiting my turn for a fare. One eventfully arrived and I took the fare to a house in a small terraced cul-de-sac some 8 miles away. I then returned to the station without picking up another fare, waited again for some thirty minutes and picked up my next fare. Amazingly he wanted to go to the house next door to my last drop. Within the radius I worked there were at least a hundred thousand options. A bit like winning the lottery you might add! Yes it is, but the difference is that winning the lottery is a very rare event but these events we call coincidental appear to happen far too often to be put down to mere chance.

When you reach the point where you realise there is no other concept you can use to replace 'coincidence' then, like me, your interest will turn into a fascination. The problem is that the more you think about it the more the word begins to hang like a blank space. I am now aware of things happening almost daily. Just last night a friend of mine told me how he could not find a banana he had kept for his lunch on the way home from his holiday trip. On reaching his car close to the airport he found one, in good condition, lying on the pavement. Another early influence on my thinking was a story by my partner. Her brother went for a short holiday in Scotland. At the time he had very little correspondence with his parents and had not told them of his attentions. They had not told their children of their own plans either, yet on a summer's day on beach some two hundred miles from where they lived they coincidently met up. What further fascinates the mind is that on reaching the long sandy beach they had a choice of going left or right and could have still missed each other.

The amount of times luck, chance, or coincidence plays a part in our lives is truly remarkable and I bet everyone reading this has similar tales to tell. Of course it can be argued that each of the incidents I have mentioned can be explained rationally, either through factors we are aware of cohering around the event, or simply a coincidence in which factors we are, or are not, aware of, cohere. I am also sure it can be demonstrated mathematically that many, if not all of these coincidences can be accounted for but I must point out you can do the same with an apple falling to the ground but it still doesn't tell you what gravity is!

Like everyone else I have always assumed there is always an explanation for a coincidence and the fact that we can work it out means there is nothing else at work when they occur. Whatever, we only have the one word for all these events, 'coincidence'. This is true whether or not the event is common or what I have termed extraordinary. I am arguing that precisely because we can explain the less mysterious events this way that we look no further when confronted with other more incredible coincidences.

In short a rational explanation is possible only some of the time the rest of the time we suspend our incredulity and call it coincidence.

We accept that Coincidence is to some extent highly predictable, especially with small numbers. The example of always finding somebody with the same birthday in a room with over twenty people in it is often given as evidence of this. We get the same predictability with a spinning coin. In other words coincidence is fixed into the order of things as a probability and as such is part of the given, a fundamental of reality. It is only when a coincidence flies against mathematical probability that we stop and raise a quizzical eyebrow. It is time to look at that reality and in the process examine what we mean by probability.

Before continuing I must issue a warning. Once you start thinking about strange happenings, especially coincidences, you will become aware of them all the time. This happened to Arthur Koestler when he was doing some research for his book, The Roots of Coincidence. (1) In the book he claims that at the time he was subjected to a 'meteor shower of coincidences'. In your own life you will find lots of seemingly minor incidents occurring; although generally minor I will nonetheless be arguing that some of them are far more significant than we realise. You can start making notes from today and I guarantee you will be pleasantly surprised. However, you must also bear in mind that as you become more and more sensitised to the issue of coincidence you will have to learn to discriminate between the mundane and the interesting and the banal and the fantastic. The purpose of this book is to point out that all these events have more significance than we realise, and to provide the route to a possible explanation.

I must also add that the perspective being pursued here is based on the notion that the sub-atomic world founded on the notion of the materiality of the atom is not and cannot be the basis of the fabric of the universe. If it were so we would be left with the dead end of the Big Bang as being the only explanation we have in our search for an understanding. In other words there was nothing and then along came something. Materially that is all we have and until we make a leap in our thinking that is where, just like goldfish in a pond, we will stay. So this book is about the search for the missing route to the hidden path, not, at this stage of our journey, the answer. It is just a possible way forward which may illuminate and add to our understanding.

CHAPTER TWO.

OTHER WRITERS AND EXPERIENCES.

The incidents mentioned so far are not the only examples from my experiences. In fact since I have started to keep a record of events they have become increasingly more common. Sometimes things happen which make us catch our breath and these are the coincidences I call extraordinary. One happened last year when I went for a swim. It was in North Cyprus in a very remote area near the Karmas peninsula – it's an area which until recently, had no road just a dirt track. While sitting and dreaming in the summer heat I chanced upon a young girl learning to swim and, being a swimmer myself, I commented that the child would soon be swimming for England. The woman in charge turned out to be the grandmother and replied in a deep, rich Welsh accent, that she too was impressed. In order to ensure she was not wounded by my reference to England I told her my sister lived in Wales in a town called Pontypridd, the pronunciation of which defines you as either a true outsider or somebody with at least a familiarity with the culture.

"I live their too" she said.

I pointed out it was actually a village above the town called Church Town.

"I live their too" came the reply.

"Do you know where about your sister lives?" she asked.

"I can't remember the name but it's on the left as you reach the top of the very steep incline."

"I know that turning and that's where I live. Do you know the name of the street?" she asked.

"No. But I remember it's a crescent."

"There is only one crescent and it's Oakfield Crescent!"

"That's it!" I replied.

"What number?" she then asked

Again I could not remember and then, delivering the coup de grace she asked,

"Is her name Betty Woods?"

I nearly fell into the water, especially when she added that she lived two doors away from my sister. That all happened on holiday and I am still amazed at the strangeness of it all.

That was strange enough but spookier still the same type of coincidence occurred again the next year. This time I was swimming at my holiday apartment. I had done my customary 10 lengths and had stopped for a rest. A couple who I had already exchanged pleasantries with and had established where both Scottish, were sitting alongside the pool.

"How long are you out for?" I asked.

"Two weeks, they replied together.

We chatted for a while and then, once again I could not resist the opportunity to demonstrate my expansive knowledge of the world.

"Where about in Scotland?" I asked.

Then followed the same verbal exchange as the previous year. My face must have mirrored theirs as we sat in awe of our shared knowledge of our parallel universes.

It turned out she knew my son-in-law, knew his occupation and was also a social worker in a town where my daughter was also a social worker, she knew my grandchildren, and was able to comment on my daughters skills – which, as a father, I found commendable. He also knew my daughter but, being born and bred in Shettleston in Glasgow, knew my grandson better and was able to tell me all about his political tendencies. And finally, to ensure we all marvelled at this massive coincidence he pointed out that the owner of the flat they were staying in was my son-in-law's old school mate and was in the same class as him throughout his youth.

What makes these more extraordinary for me is that they happen in a remote area of the world. It's a place I would expect to see few people from my past – let alone people who know my family. It's in North Cyprus a country unrecognised by the rest of the world apart from Turkey. It has a population of about two hundred thousand. If these coincidences had happened on a visit to the Isle of Man where about twenty members of my family live I would not be surprised. I must also add that as a teacher for thirty years in the same school in my home town of Liverpool I never, ever, met more than four or five people out of the thousands who I had either taught or who went to the same school as me over the last forty years.

I was curious to see what other coincidences were most talked about and went on the internet to find out. This led me to a long list of books. There are an enormous number of stories and I include but a few. One of my favourites is from Coincidence or Destiny by Phil Cousineau. (1) He relates a story about Anthony Hopkins who unable to find a book about a film he was to play in. Coincidently he finds one which had been signed by the author, stolen and then left on a bench in the train station. There are many other amazing coincidences in the book and it is well worth a read, as is The Power of Coincidence by Frank Joseph he relates a story from a book by Warren Weaver where fifteen members of a choir were saved from certain death because they were all, for 'trifling reasons' late for a meeting. (2)

Out of the large number of books I have read on the subject of coincidence I have mentioned only a few. In fact once you start getting interested you will find that there is a mass of literature on the subject. And probably the most important contribution came from Carl Gustav Jung when he coined the concept of 'Synchronicity'. This is the idea that there is a connection of some sort between two or more events and that this is unlikely to be due to chance.

Carl Jung coined the phrase following a session with a young woman he was treating. She had a strange dream which appeared to anticipate the emergence of a Scarab Beetle at the window. An event highly unusual in his locality. (3)

Coincidence is all about us and many examples and anecdotes have been comprehensively documented by a large collection of writers. However one of the first to try and produce an explanation for these events was Paul Kammerer an Austrian Biologist who collected his observations, published a book about them and used them to formulate a theory of what he called 'Seriality'. Basically he too argued that there were forces in the world which accounted for coincidences and that these events were connected. (4)

Kammerer believed coincidence was embedded in our culture and that culture does not operate in a straight line. He believed, that we live in a population or a general grouping which means that there are parallel or recurrent themes which take place across society and account for what we perceive as coincidence. It is based on the idea that culture cocoons and constrains our thinking, especially for what we perceive as rational. Within the confines of this world our behaviour patterns reflect that grouping. When we step outside the parameters of this population we cross a threshold and that's when stranger coincidences occur.

He made an enormous amount of observations and would even count the numbers of passers by carrying umbrellas. Einstein is reported to have found his ideas interesting. You may want to argue he was just being diplomatic, but I would point out that Einstein was fully aware of the possibility of 'spooky action at a distance', even though at that time he was busy trying to refute the possibility. 'God does not play dice', he said.

Kammerer identified seventeen categories of coincidences. He too was clearly of the opinion that there are certainly very different types of coincidence and they appear frequently enough in our lives to be of greater significance than to be dismissed as chance or luck of the draw.

I must certainly agree with Kammerer that culture plays a significant part in our consciousness. We are social animals and without a culture there is evidence to show we could not speak, walk or carry out the most basic of tasks. Culture is needed not just for dealing with what we relate to through our senses, but also our imaginations.

The role that the consciousness plays is very important. Note that in all the cases mentioned it does not become a coincidence until we think something is going to happen either consciously or unconsciously. This is clear from the example taken from Carl Jung. He states that at the time he was thinking about what he could do to shake his client out of her mood. Our expectations, predictions, hunches are inevitable but they are unavoidably part of our consciousness and the process of categorising what is or what is not a coincidence. It would be a coincidence now if a hedgehog appeared in my garden as I am now thinking about one just for the purpose of giving an example of what I mean. I would not call it a coincidence if it was a cat as they are always seeking to get in when I am not looking and I am always conscious of that possibility.

In most of the cases I have mentioned it is clear that the consciousness plays a part if only that it is a marker for a coincidence. In other cases it is not as clear. However, we have to recognise that there are no coincidences without thought. Thought is not separate from creation but is a part of creation whether the breath of God or the result of the big bang. A coincidence without thought being involved as a pre-curser to, or participant in, is not possible. Imagine that on the other side of the moon there are two rocks with strange writings on lying nearby each other and shaped like the Rosetta stone. Until discovered by an astronaut and given up to consciousness they are meaningless events in the existence of the universe. We cannot avoid the role of consciousness especially as in later chapters we have to confront one of the more amazing perspectives of quantum theory which is when we look we change things.

You can get involved in the age old philosophical debate about, cause and effect or space and time, but while these concepts do help to shape our experiences they do not limit the role of consciousness. In fact the modern world of physics works on the basis that you cannot have space without something being there, like an object or, as we shall discuss later, a field. We know the material world is there because we can feel it through all our senses, but we cannot feel empty space. We cannot even think of an empty space; we can only think of it against something else having something in it. In order to picture it we would at the very least give it a property like colour and then depend on our ability to make a leap towards comprehension.

Consciousness maybe a mystery but it is also a tool which transcends our predispositions. Think of a cup you are drinking out of. It will change its shape constantly as you use it but you will, no matter what, contain the concept of a cup. This does not mean you cannot step outside and put aside its limitations on thought. Science, very recently, has now discovered that we do the same with colour. They have learnt that whatever the colour light reveals an object in we will retain that colour even when the light changes. Importantly they also point out this too can be changed by other experiences, meaning that despite the paradigms we use to interpret the world they are not fixed in or set in stone – as this discovery proves. In effect my black and gold may be your black and white. This does not mean we are constrained to see reality only the way nature decides as we not only have many different ways of seeing things but also have that very special aspect we also do not yet understand called consciousness and its aspects of comprehension, intuition and bloody mindedness.

You may argue that I am merely identifying the role of the subconscious in the events we call coincidence. That is exactly what I am doing as it is my belief that the discoveries in the world of science have a greater significance than we yet realise in understanding the reality of existence.

The last few centuries have been centred on the search for an understanding as determined by the world of science. The answer is deemed to be a neat mathematical model in which all the parts fit together like a jigsaw. The perspective, generally accepted, is that we proceed through experiments which can be tested and then either proved or refuted. There is no room for any other approach as outside the laws of science lay the insane and ideological world of religion and the mystical. Well, the part played by science is itself now under the microscope and in the next few chapters I will elaborate as best I can about the divisions and strange theories within the world of science.

I will not be talking about coincidences again until we have looked at the latest observations in the world of science so before moving on I want to relate my latest experience. All coincidences are not the same and sometimes they can open up our minds and act as a stimulus for thought, bring many events into focus, or remind us of how much we are at the mercy of events outside of our control. The other day I go to the dentist as the tooth next to the one he root-filled is also now in need of a root filling – two months after the last filling. I have a different explanation than a coincidence, but it may well be just that? I lean back and he starts chatting. He always relaxes me, so much so I am feeling quite blasé about it. 'Have you read this book?' he asks me. It's a UCATT book not in the public domain, he adds, and it all about Robert Tressel, author of The Ragged Trousered Philanthropist. I have just that morning read an e-mail about a recent lecture by Jon Cruddas MP all about the Author. Surely two people totally unrelated, and miles apart bring up a subject without any prompting by me, is not a coincidence. Its grist to the mill so I tell him about the book I am writing. It's all about the way in which quantum affects our everyday lives and accounts for such things we call coincidences. He tells me about a film called Magnolia as he too is interested in coincidences, and relates some of the strange stories.

In short we swop stories. I tell him that somewhere in a parallel universe I may be pulling his teeth out. Oh, we chatted like good old friends for ages, In fact the cocaine nearly wore off. And then he gave me the bill £255. Nothing coincidental about that – well not in this universe.

On the way home I couldn't help thinking about it. My imagination, that thing we call consciousness, set to work. What if events in one dimension have an effect on events in another? A stream of incredible ideas passes before me all making no sense whatsoever. Eventually I decide there may be a multitude of dimensions, possibly an infinity in which we can assume an almost incomprehensible flux like, coming and going. This pleases me as it means that somebody somewhere hasn't got this toothache and the sparrow killed by the sparrow hawk in my garden this morning is chirping away happily in another dimension.

What is being argued here and in many other books is that there is more complexity to life than we know. Somewhere in the great scheme of things there is more significance to what we call a coincidence. It is not simply a one bottoms up universe all built on our idea of a mathematical formula. That is the Goldfish view. It stems from a conceit about our ability to know everything and it lacks the humility of the view that we are only just beginning to unravel the mystery.

I believe we need to once again see our lives as a web of moving parts in which the bits we see are like the shadows from the clouds or the pixels on a TV screen. Nowhere is there a better understanding of this than in many areas of the world of science and the incredible ideas produced almost daily in the search for a greater understanding

My brother recently told me a tale from when he was younger. He shot across a junction in his car and realised he had not seen a car speeding towards him. He accepted that he had caused a fatality and all went black and quiet. But then everything came back with a roar as he realised that he had cheated death. On hearing this story his wife remarked on it, calling it a 'leap across time'.

I am telling the chap next door about this event and he then starts telling me about similar happenings in his life. He eventually goes silent and mutters weird man, really weird. And that makes me decide to start writing things down, in the hope I can pass on what little knowledge I have of the way things may or may not work. However, before I can develop my arguments further we need to come to terms fully with what is happening in the world of physics and in particular what is meant by 'Spooky action at a distance.'

Maybe there is something in the books I read that will help so I am reading Steve Bryson's excellent book Called 'A Short History of Nearly Everything' and have just finished the page where he writes about the discovery of a new species of our early ancestors in Neander Valley in Germany which we now refer to as Neanderthal Man. He then points out another coincidence - the word Neander in Greek means 'new man'.

CHAPTER THREE
THE GENERAL PICTURE. AN OVERVIEW.

It is my belief that coincidence and order are part of the same thing – they belong to the same family. Coincidence, just like cause and effect, is treated as common sense. It is the rational explanation and the only explanation we have for regular events in our lives. However, cause and effect are two events in time, the effect occurring after the cause. It is the brain which treats it as a single event but, very importantly, can also comprehend its role in thought. We put these two events together without ever questioning the mental process. This leads some to argue that what we call common sense can also be seen as a built-in program, a precondition of our being through which we can interpret the world. I believe it is possible that 'coincidence' is very similar, a precondition, a device to explain things we do not yet understand. Like cause and effect coincidence helps account for the relationship between two seemingly unrelated but parallel events. It provides a rationalization which is also part of our basic programming. This does not mean that we are like the goldfish who can never see further than the within the limits of his pond. We have consciousness and can recognize the issues and step outside them. However, I do believe that only an active unrestrained imagination can possibly take us any further than the limits of our pond.

For coincidence to occur and be explainable through something different than probability we have to play with the idea that it is a parallel event and that there are possibly other dimensions we do not know about. If there are, as you will find many much respected theories propose, different or other dimensions, then it would mean that we should not be too surprised that coincidence is currently explainable only through probability theory.

We are all familiar with the way probability is used to explain that if you toss a coin often enough the chance of getting heads or tails will turn out to be fifty percent. Probability means exactly what it says – it is not definite, only probable which means it may not happen like that. In the mysterious world of quantum mechanics, which we will discuss later, they have a mathematical formula relating to the square of the height of the wave which tells them the probability of finding where a particle might turn up but cannot guarantee that it will, which is like saying its over there somewhere. Another way of understanding this idea is to imagine a swimming pool rippling with David Hockney waves. The wave is spread out across the pool and in it beneath the water is an underwater swimmer who is going to come up to breath. Predicting where he will come up to the surface is made more difficult by the fact that you do not know how long he can stay down. However, the task can be made easier if you know which pool the swimmer may be in, and you have a mathematical formula which helps you make a fairly good estimation - probable, likely, maybe and maybe not!

If there was no probability, if everything was predictable, we would be very suspicious and could safely assume there was only one dimension and everything could be reduced to a mathematical certainty. We now know that is not the case in our universe. Mathematics is logical – nature, more specifically, the quantum world, is not!

The thought that there are other dimensions is an easy one to make as it feels intuitively right, fits in with the idea of coincidence and as you will discover later is an idea readily accepted by many of the greatest thinkers in the world of science. That may simply be because nobody can accept the idea of an infinity, a going on forever and ever of one single universe. But it is also difficult to accept the idea of a single dimension as it follows that if you have a dimension you must also have what is outside the dimension. And, that has to be either a new infinity or another dimension. Some theories, like string theory, have used mathematics to claim they have actually worked out how many dimensions there are and we shall look at them later too.

I must point out here that what I am about to undertake over the next few chapters is an explanation of what is going on in the world of science. This is necessary if we are to arrive at a theory about coincidence which is based on the evidence available over the recent past.

That will be seen as an audacious and seemingly impossible task as there are very few people who can understand the scientific jargon of physicists, which you find on most websites and in books that are dealing with the particle world. This is not only true for people in general but also for scientist involved in different disciplines, like biology. If you look at the jargon used in the world of physics and the particle world it may look like they know what is happening but if you ask them they, including every top scientist who ever lived, will all truthfully tell you they do not know why! That leaves us free to shine a philosophical light on some of the issues. That is my excuse for taking on the task of developing a basic understanding.

I originally wrote this for myself but have since decided that as the topic is becoming more and more accessible to the public others may find some of the information useful. I also want to point out that there is no simple way of bringing all this complicated information together so I have broken it down in a way that is partly repetitive but hopefully educational.

This morning the 9th February 2014 I read the following on my Yahoo news page. *"Exotic particles never before detected and possibly teensy extra dimensions may be awaiting discovery, says a physicist, adding that those searching for such newbies should keep an open mind and consider all possibilities.*

Such particles are thought to fill gaps in, and extend, the reigning theory of particle physics, (The Standard Model) said David Charlton of the University of Birmingham in the United Kingdom, who is also a spokesperson of the ATLAS experiment at the world's biggest particle accelerator, the Large Hadron Collider (LHC), and one of the experiments thought to explain why other particles have mass. (The Higgs Boson)

Physicists should not simply be searching for evidence to support one theory or another.....it is important to look at every rare process we can that might be a signal for some new physics showing up.....We really have to try to be as open as possible and try to leave no stone unturned in looking at all the possibilities," said Charlton. (1)

Great news I thought, we are not alone! At last there is real movement in the world of science so there is no reason to write this book. Then I remembered I am writing it both for my own understanding and also to make these findings and new theories accessible to others. That is what this book is about, but first I must explain why I believe my comments are of any significance, what my thoughts on scientific statements about reality are in general and then outline the research findings and other issues we are going to examine more closely in future chapters.

My interest started a long time ago when I was a young man. That was in 1965. I mention it because at the time I was working as a milkman and would deliver to the university. Thanks to a few friendly students this gave me access to the latest issues and experiments. One experiment in particular fascinated me. It is known as the two slit experiment and has had massive implications for the world of science.

The debate began around the same time in which it was realised that light was able to act as both a particle and a wave. We are hearing the debate now thanks to the widespread growth of the media, in particular the World Wide Web which makes this knowledge and related experiments accessible to all who want to see the latest news, available at the press of a button. It is also a precondition for scientific research grants that both the government and the public are kept informed of all the latest developments in science. You can drop what you are doing right now and go and have a look. In fact everything you read here can be confirmed, questioned and discussed through the internet. Hundreds of sites will give you all the information you need including a video of the top scientist in the world elaborating on their work. If that is not enough then obtain a copy of John Gribbin's excellent book Schrodingers Kittens, David Deutsch's book The Fabric of Reality or Richard Feynmans book The Strange Theory of Light and Matter. You will find them all referenced at the back of the book.

Scientists and philosophers have been aware of most of the research we will talk about for a long time. However, despite enormous progress in all our lifetimes, they are still unable to answer the most fundamental of questions. In fact there is a continuous and ongoing debate which has still not been resolved and many believe we will probably never understand. Among them is the president of the Royal Society,

Lord Rees argues that: *"Some of the greatest mysteries of the universe may never be resolved because they are beyond human comprehension, According to Lord Rees, president of the Royal Society. Rees suggests that the inherent intellectual limitations of humanity mean we may never resolve questions such as the existence of parallel universes, the cause of the big bang, or the nature of our own consciousness....A 'true' fundamental theory of the universe may exist but could be just be too* hard *for human brains to grasp,"* The Sunday Times June13 2010

Having other interests and hobbies I believe is an advantage as I arrive uncluttered with the minutia of the debate so far and free to concentrate on the wider picture. I think this is important because the majority of us are simply the recipients of whatever the world of science, or the media in general, dictates to us are the facts. Hopefully I have enough of an understanding and am widely enough read to say something which will appeal to those others who are interested in the big question but like me do not wallow in a scientific background.

I am not questioning the fact that we are able to manipulate the particle world and also believe the knowledge we have accumulated over the last century has been incredible and has allowed us to reach a point where it is clear to everybody involved that we are only at the beginning of an understanding. It follows from this that trying to prove that there is something more to coincidence than the commonly accepted view is not possible without recourse to the very latest results of scientific experiments. And, as Science is where our current knowledge of reality is largely driven then that is where we need to go to find out what is going on.

We need to start by pointing out that when we say something is scientifically proven we are actually talking about a methodology, meaning it has been researched many times, in a particular way, and the results are always the same. The fundamental rule for establishing your theory as a fact is that you can set up an experiment so that it can be proved or disproved time and time again. This methodology has implications which we will look at further on. It is also important to state at this point that I believe that to see science as the only way of arriving at the truth is a nice idea, and very desirable, but there is more to it than that. We need to remind ourselves that it is the leap of imagination that takes us forward and it is the scientific method which makes up the dialectic process. In other words the truth, if it is ever to be known, comes out of the process between the two, imagination and research.

Science, like every human construct, is an ideology which presides over all scientific judgments about reality and we need to be aware of this once we put our thinking caps on. Ridicule is often heaped on anybody not accepting the scientific view, a view often promoted through the media by people whose prime motivation is to promote themselves or their own personal program. Sometimes this is done through a jargon used to beat off any criticism by wrapping the information in a private code or with the comment it can only really be understood by a mathematician. We need to remember that mathematics, like science, is just another means of communication. Both have their own language and just like any other language they will be loaded with their own cultural baggage. Having said that I must add that I do not believe it is sensible to be anti-mathematics or anti-science and so I intend that any future speculation needs to be based on the firmest foundations possible and the best way to achieve that is by examining the very latest findings in the world of science in order to obtain a greater understanding.

Such an approach creates the need to elaborate further that we cannot ignore the influence of ideology and for that reason I want to talk about the work of a number of great thinkers, among them, Thomas Kuhn, a physicist who made a significant observation about the way in which scientific discovery proceeds. (2) He pointed out that if you study its history closely you will find that discoveries in any particular area depend very much on our mental map about the ideas of the time. These ideas he likens to the form of a paradigm which serves to steer the theory in a particular direction. Any attempt to move outside this paradigm is not only frowned on and discouraged but also brings about a loss of status, respect and loss of credibility. This is very much how many people now see the current push to find the missing piece in the puzzle. Any indication that the currently accepted idea of what is deemed a sensible and scientifically approved idea is being challenged and you are in trouble. If you do not lose your job you will certainly undermine your status and credibility. Generally any alternative theories are only going to be taken seriously after all avenues have been exhausted and enough data or challenges have been unearthed to allow for change takes place and a new direction to be considered. That's called progress but at other times a revolutionary change takes place thanks to somebody being able to think outside the box and having the independence of mind to be undaunted by the establishment.

Karl Pooper another great British philosopher had clear ideas about how such revolutionary change took place and he saw science as in the business of refutation, meaning the job was to prove something wrong, mistaken. (3) He argued that change occurs when the standard theory or the theory of the time gets unmanageable and collapses and a new theory takes its place. Kuhn adds another dimension to this argument by pointing out that the social norms in the world of science are no different than in any other walk of life and they determine what progress is being made. Put these two ideas together and it is easy to see how the current ideology or value system of the period can actually hold up development. This is not new and is and always will be an issue for scientific development. This has happened in the past and will happen again in the future to people who offer alternative theories. Galileo is the obvious example. He was prevented from revealing the fact that the earth was not the centre of the universe.

I would agree that things have moved on since the Dark Ages and are destined to move even faster in the future now that we all have access to the world of science and its wonderful challenges. That's why I am able to write this book, offer my own interpretation and make it accessible to a new digital world of information which will be available to everybody who wants to get involved – if I can find out how to upload an ebook!

It is because of the wonderful up-to-date information we get from the media that I am able to start my discussion of science with the very latest revelations. In order to grasp the significance of this I would point to a TV programme in which I saw one key modern day figure in the world of science, professor Max Tegmark on a TV show talking about the time he felt he had wasted learning all the names of the numerous particles. The idea that everything is made up from very tiny particles called atoms has been around for a long time and even at the beginning of this century they were aware of up to a 100 particles but since then they have continued to discover a whole range of new ones and there are now so many they call it a particle zoo. He then went on to talk about the 'Standard Model', which many believe is almost complete, as they just need to know where mass comes from. Without mass, particles wouldn't hold together and there would be no matter. I have heard it described metaphorically by a distinguished professor as being made up of six loaves, six tomatoes, five partridge and God. We must also continue to bear in mind that the Standard

Model does not account for gravity and other forces, among them, Dark Matter and Dark Energy.

We will discuss the Standard Model after we have looked at a few of the basics. For this chapter, and the rest of the book, you need no more knowledge than the following. Atoms are made up of three basic parts, protons, neutrons and electrons. One has a positive charge the other a negative and one has no charge at all. Think of a charge being a magnetic type pull or push. Two of these bits stay put in the middle while the third, the electron, whips around them here there and everywhere. Whether the atom is gold copper or lead depends on the combinations of protons and neutrons. We also need to add the fact that the middle bit or nucleus is very small compared to the whole atom. It is about the size of a pinhead in the middle of a football stadium – a lot of empty space! If you join up a number of atoms, and their empty spaces, you will have a molecule or family of atoms making up the substance of our universe. If you want to know more you will get all your questions answered on the World Wide Web.

For some time they have been smashing these particles into each other. According to the BBC they make ten thousand collisions a second in the search to understand all the particles and discover any new ones. It was this process which made them aware of the particle zoo and the same process which led to the confirmation of a theory of some new particles they call quarks. It was only in 1995 that some believed they had found the last one, the TOP quark. There are very few now who believe they have all the pieces making up reality. They do not even know why there are so many particles.

Nothing I have written so far would be a surprise to a modern day scientist. They are fully aware that we still have a long way to go and that there are still a lot of unanswered questions. There is now a mass of TV programmes littered with famous names of highly esteemed scientists working on a wide selection of alternative theories. Many, if not all, will look straight into the camera and tell you that they do not yet know the answer to some of the most fundamental questions.

This is especially true of what we know as quantum theory. There is no simple explanation and there is no complicated explanation as nobody understands it. They only know that it works as a scientific tool. It appears that everything in the sub atomic world can behave as both a wave and a particle, communication is through waves, and you cannot predict with certainty the position or direction the particle is going to pop up in. We now know that the act of trying to measure the sub atomic world in some way means you end up with a particle but if you do not measure you may end up with a wave. There is no certainty in anything you do and all equations are at the mercy of probability.

All of what is written here has been confirmed through scientific experiment and can be validated through any scientific journal which you can get to through an internet connection. In order to develop an understanding of the sub-atomic world and its strange, incredible and mysterious behaviour we need to have some understanding of quantum mechanics, which is sometimes called quantum physics or quantum theory. The short version of quantum is that it is an acceptance that the relationship between matter and energy is both as a particle and as a wave.

But it is much more complicated and amazing than that and we can start delving into it with a few examples from the sub-atomic world by reading the following quote from Richard Feynmans book, The Strange Theory of Light and Matter: " *We must accept some very bizarre behaviour: the amplification and suppression of probabilities, light reflecting from all parts of a mirror, light travelling in paths other than a straight line, photons going faster or slower than the conventional speed of light, electrons going backward in time, photons suddenly disintegrating into a positron-electron pair, and so on*" (4)

Quantum mechanics ought to be explainable when looking at how it works – but it isn't! They can work out what is happening but have not yet figured out why. This is evident when looking at the laser. The atom has a nucleus, a complex arrangement of parts, among them protons, neutrons and electrons. When an atom is hit with a photon of light one of its parts, the electron, gets its balance disturbed. In other words you start with a happy family of atoms all sitting and singing to each other. You then start messing their life up and get them all excited. This speeds up their life cycle and some start giving off light energy, that is change into another form, and emit radiation or waves in the process. The radiation wave is like a disease and spreads to other atoms and they too start to do a dance. Now the family is unstable for a few milliseconds but before the equilibrium is restored the electron shoots off the surplus energy caused by the hit. However, for some reason, to do with probability, it produces two photons instead of the one. The situation is now snowballing and you have them all trapped in a tube. If you have a few mirrors at each end you increase the spread of the radiation which creates even more panic as it bounces back and forth and multiplies the number of photons. This energy is trapped and with the aid of mirrors multiplied millions of time so that you get a beam of light all of the same wave length and with much more intense than ordinary light. If you knew exactly how this works you would understand quantum mechanics and be able to explain fully the next paragraph. Nobody does so it's the best non-scientific explanation you will get as to how lasers,

TV mobile phones and a host of modern appliance work. (5)

Another more impressive explanation about the mystery of the amazing goings on in the sub-atomic world is provided by Jim Al-khalili and JohnjoeMcFadden in their book Life on the Edge. From the very beginning they discuss the notion of tunnelling, which we will look at later. Tunnelling is a mind blowing theory and in the book it is pointed out that that 'particles walk through walls'. (6) A footnote at the bottom of the page alerts you to the fact that they do not really know what is going on but the analogy is mind blowing. It is the reason the sun is able to shine as some particles fuse together to make energy and to do this they have to work against forces that try to keep them apart. They do this by using their trick of becoming a wave at the right moment and then fusing with another particle before it knows what hit it. This is worth reading and it is explained more fully in the opening chapter.

Quantum mechanics is so abstract and runs against our common sense notions that it is incomprehensible and understood by nobody, and that includes mathematicians. You might think it can be explained through mathematics. You would be wrong. You have a choice, mathematics is never going to root out the truth on its own or quantum theory is fundamentally wrong. Very few believe the latter.

We are now at an impasse and only when we produce a new idea will mathematics have a role to play. The perspective promoted here is that the reality of our everyday beliefs is shaped by the mind and the information coming in is shaped, organized and made sense of within the paradigms and ideology we are steeped in. Consequently, in order to dig deeper into the fabric of the universe we must be prepared to break out and entertain other ideas. Intuition, luck, coincidence chance, are all there to be used in our search of the truth.

This is not to ignore the role of mathematics in providing a formula about what they call the wave function, a mathematical formula about speed, momentum, uncertainty and probability I propose the following perspective on Mathematics. Almost everything we think we know is, or once was, a mathematical theory, a theory which has then channelled research in a particular direction in order to confirm that the maths was right. It is the natural thing to do, more so if, like some, you believe that maths mirrors reality. But not everybody believes that maths mirrors the underlying reality, rather we may be wasting a lot of time looking in the wrong direction. The good news is that with the emergence of quantum theory and its implications many Mathematicians have already been asking the same questions. In all cases mathematics can be seen as a tool, just like in a computer, which we are able to interpret reality through. It is not reality but it is a way of ordering reality As such it cannot be separated from the world out there. And, without intuition, comprehension, guess-work, refusal to follow the crowd, then Mathematics, like any other ideology, is capable of becoming more about the network than the world out there. Current research into the workings of our brains indicate a deeper and unconscious activity which bears no relationship to everyday experience. It's an understanding of the necessity to think outside the box. It is also a belief in the mysterious, a mystery we are currently confronted by which is epitomized through quantum theory. Mathematics is a tool we consciously use to help us understand the world out there. It is not a true reflection of the world out there. Just like we invented the hammer and the nail we invented the zero.

Quantum theory is extremely difficult, if not presently impossible, to understand but as everybody, Einstein and the great American Scientist Richard Feynman among others, tells you nobody truly understands anyway. Like most people we have to remain in awe of what is happening in the subterranean world. Let me repeat, nobody on the planet earth understands what is happening. It is a complete mystery and we can only guess at the answers. We cannot predict anything with certainty. We can only say it is probable or likely to happen. It also means that it may not happen or the improbable may happen. Things in the sub atomic world are neither one thing nor the other, meaning they exist in a superposition of all possible states until we, or the rest of the universe observe them in some way. In other words the particle is everywhere until we pick it out.

That's the general picture and hopefully you are interested enough to look at things in more detail. The following chapters will look at the Two Slit Experiment, the Standard Model, and the latest findings from the 'Atom Smasher' at CERN, the mysterious behaviour of the particle world, its implications for the human brain, the Big Bang, quantum Theory and time.

CHAPTER FOUR
THE TWO SLITS EXPERIMENT.

Quantum theory is now seen as the most awe-inspiring hypothesis in the world today. At one level we know it works as a theory because it accounts for so much of the world we live in today, TVs, lasers, and mobile phones would not be possible without taking advantage of the strange work of the world of particles. That means we have to explain and predict events in the sub-atomic world as being the result sometimes of waves and sometimes of particles, and just to make it a little more complicated we cannot tell with any certainty where the particle is going to pop up. With all these items we are dealing with huge numbers of particles which would have no effect unless they behaved with some probability. It's a bit like saying we know that when we throw things in the water we learn which ones float and which ones sink and that enables us to build boats, but we don't understand fully why it happens. In addition if they do decide to float we cannot tell where they are going to surface.

At another level it tells us that we may be totally and absolutely wrong about everything we think we know and need to start all over again if we are to ever understand the nature of reality. If Lord Rees is correct we may not be able to do so. Quantum theory indicates there are rules or laws we have no understanding of, especially when it comes to predicting the behaviour of minute particles or entities. And let's not forget that involves everything in the universe, as everything is made up of waves or particles anyway.

It has now reached a point in the hunt for a comprehensive theory that many of our great thinkers are arguing that there are possibly an innumerable number of dimensions existing in a world of 'probability waves' and have no difficulty claiming that we may one day glide through wormholes within the Universe to look around other cosmoses or travel backward in time. In fact I was listening to a Radio 4 programme in which an American scientist was describing the possibility of multiple universes and how man has gone from understanding himself as living of a world as the centre of God's creation to one revolving around the sun and then being in a galaxy and in a universe and now having to consider the possibility of multiple universes, each understanding making man more marginal. He pointed out that if there was an infinity of happenings it would mean that an event like shuffling a pack of cards identically is bound to be happening somewhere because there is only a finite way of the cards being shuffled, however enormous the odds are – likewise the possibility of two cars arriving at the same spot on the narrow bridge at the same time.

I make you aware of these very different views in order to create an understanding that the door is wide open and it is the perfect time in history to open our minds to other ideas – ideas which have been cast aside for far too long, ideas without which we will never begin to understand the brilliance and beauty of the universe in and about us.

What you may ask is the cause of all this speculation and bewilderment. Any understanding must start with the two slits experiment. You can read almost any book on quantum theory and read about this experiment. I recommend John Gribbin's, Schrodinger's Kittens and the Search for Reality.

This one chapter I am about to write will not bring home the full implications of the research in this area so I suggest the reader go on the net and just type in 'The two slit experiment' and you will get actual demonstrations. (1)

I will try to explain this further as best I can and I am sure it will leave you with many questions. But I can assure you that if you study the literature you will find that all avenues have been explored and you will have to face up to the sheer incredibility of the discovery: light acts as both a wave and a particle and our understanding of what is going has led to some amazing theories.

We know what happens when you throw a pebble into the water, apart from, a bouncing pebble you get waves. So it will not be too difficult to grasp that if you create a wave which travels through two holes or gaps in a wall that when the wave comes through the other side the waves will actually interfere with each other. Some will cancel each other out others will add themselves on to make even bigger waves. If you direct a body of water towards two holes or slits in a wall you will be able to see the effect on the other side. It is very clear from above. The two holes have a wave emerging and bumping into each other and either cancelling each other out or making another wave. The pattern of waves from above is very clear. Now, with a stretch of the imagination, imagine we can reduce the water to little globules. If we now direct them at the two holes one globule at a time we would expect to obtain no waves and two different volumes of water on the other side.

We would be amazed if we still got the wave pattern but let us suppose we did get the wave pattern. The immediate question is how do the globules of water know which slit to go through to produce the pattern? You may think the solution is to look at which slit each globule goes through. The problem is that when you look it does not happen. The wave pattern disappears.

Light behaves in the same way. You shine a light through two holes or slits in a wall and when they get to the other side they interfere with each other in the same way, although this time you get light and dark stripes on a screen. In fact that is why we know that light is able to behave as a wave – through experiments.

However, we also know that light can behave as a particle – first discovered with the photon. We can reduce light to photons and send them through two slits just like we can do with water. That is when things get really weird. Imagine you are randomly aiming at two holes. If we were throwing apples or globules of frozen water then common sense tells us we should have two piles of different sizes. That is what you would expect the randomness of your shots to achieve.

That is not what happens. Amazingly, as with our water experiment, when you have finished you will find that the two piles are identical, with exactly half the apples in one, and half in the other. You would certainly think it strange and would check you had not made a mistake. That's what happens when you use photons and other particles, including atoms, and send them through the two slits one by one. Instead of getting a random scattering of dots we still get a wave pattern. No matter how many times you conduct the experiment you get an identical pattern of particles which defies logic and makes nonsense of everything we know.

That's strange enough but it gets even more challenging. If you had put a camera up to observe which hole the particles were going through during the exercise you would find that one of the piles would be smaller than the other. When you looked you changed things! There is no other conclusion, by looking with a camera you had made things happen differently to what would have happened if you had left things be. And that's exactly what happens. By looking it appears that you changed the order of things. And now we know that this same phenomenon happens not just with photons but also electrons and atoms. You should be amazed and would certainly be if apples, carrots and bars of chocolate, behaved like this.

In the scientific experiment the screen is a photographic plate and the visit of each photon is recorded. Thousands and thousands are delivered over a period of time and at the end of the experiment you can clearly see the typical striped wave pattern. And that is still happening even if you are firing each photon individually. I repeat the question: how does it know which hole to go through to make the pattern we see, or how does it go through both holes at once? This is made all the more amazing when you realize that the experiment was arranged so that they could observe which holes were used. When they did the pattern disappeared. It was also arranged that the holes could be opened or closed at random, and still the little chaps were able to work out which hole to go through. There is clearly something we do not know about light. They even did an experiment in which the scientist never decided on a whether or not they were looking for a particle or a wave until after 'it' had passed through the slits. This produced an even more incredible result as there was no difference except to suggest that 'it' could go back in time in order to change things. I can enlighten the reader no further on that.

You may want to believe it is all because there is something special about photons. The problem with that theory is that it has all been done with electrons and with the biggest daddy of all, atoms. The Electron experiment was done by a team from the Hitachi research labs in Tokyo and the results were identical to what happens with photons. It would now appear that the same laws apply to atoms and this was established by a team from the University of Konstanz in Germany in the 1990's. These experiments have probably been done may times since then as I am taking the data from chapter one of John Gribbin's book mentioned above and published 20 years ago. In short atoms aimed at two holes appear to go through both holes at once which is no different to accepting the idea that atoms appear to be able to be in two places at the same time. In fact that is exactly what scientist now believe to happen, any of the so called particles of the sub atomic world can be in at least two places at once.

You can catch up on all this information by watching any of a mass of videos now available on your computer.

The usual theory used to account for what is happening is called the Copenhagen Interpretation. Atoms, photons, electrons travel as waves and arrive as particles depending on whether or not they are observed. That's right, observed, meaning that if we are not looking then it does not exist or if it does we do not know what it exists as. In fact we cannot even be sure where it is at any moment.

It is all very spooky. The idea of spookiness is best understood through the concept of what they call 'spin'. It is a scientific term and it is the only way they can identify individual particles. Without 'spin' they all look exactly the same. If you shine a light at a crystal you can actually split a photon into two photons. However, a change talks place as each photon will then have a different polarization – called spin. Stephen Hawkins, in his book, A Brief History of Time, points out that 'spin' is a strange explanation and is not the same 'spin' we see in a top or when we turn around. It is seen more as a potential which a particle has. A dot is a dot from any angle but other particles are different from different angles and more remarkably, they can be different again if they are turned around again. One particle has to be turned around twice before it gets back to its original state. The only way I can imagine that is imagining a figure eight from side on. By turning it through like you would a train set on its track. You would need to turn it twice in order to return it to its original position. (2)

I repeat, when you split a photon in two you entangle them so that if you do something to one of them you will have an effect on the other. Amazingly there is a connection between the particles no matter how far apart they are. This is not an infrequent event as entanglement is quite common in the particle world. A particle gets taken in by another particle as it passes through an object. Then the new particle throws out the old particle but sometimes when it does this the new particle is emitted at a lower wavelength and as the total energy of the two must by law equal the energy of the original particle (we are told) then we now have two twins. However, they are not identical but entangled in some way. Einstein named this as 'spooky action at a distance'. There have been some developments in this area which we will talk about later.

Now just in case I have not yet got your full attention I will tell you of one more remarkable experiment. I was coincidently at a lecture on Black Holes, mainly because I thought it may be a useful learning experience and it may even help me with this article. It was delivered by a professor for a local university and was far too advanced for most of the audience to digest fully. However, it was fairly informative. During question time the professor told us of an experiment being conducted to try and help towards a holistic theory – one that provides us with a picture of everything. She mentioned 'cosmic foam' which was something I had never heard of. On getting home I looked it up and that was the moment I discovered the name of another great physicist, John Wheeler. John knew and worked with Albert Einstein and Niels Bohr. He has a massive reputation among physicists, and founded the name "Black Hole" to describe the dense, light-trapping objects now accepted as an important part of the universe.

Wheeler proposed a version of the two slit experiment out in space. Remember, despite sending photons through one at a time we still get a wave pattern on the photographic plate. This should not happen and defies what we know about how the physical world should behave. But it does happen, and when we try to see which hole the photons are going through we collapse the wave and loose the wave pattern. In short when we look we change things.

John Wheeler adds another perspective to the debate by proposing a thought experiment set out in space. (3) Incredibly he was suggesting that our observations in the present effect the past. Huge planets bend light and so if we imagine them in front of a bright light source like a quasar we can treat them as acting just like slits in the two slit experiment and by some clever arrangement would end up arranging things so we obtain the same wave patterns as we do in the laboratory in the two slit experiment. This implies that once we look to see which side of the star or galaxy they went around, the photons, which had been on their way to earth for the last million or so years, would behave the same way as in the two slit experiment and the wave would disappear. So, take your pick, they suddenly changed direction or the past was changed? Whatever, the experiment suggests that the paths the photons take are not decided upon until they take their measurements, even though their interference was made after the photons had already left the source of the light. The implication is astonishing for it means that we do not only affect the immediate event by looking but we also create the events history. In the words of John Wheeler, 'We have a strange inversion of the normal order of time…an unavoidable effect on what we have a right to say about the past history of the photon'.

You may now see what the fuss is about. It is simple, according to our everyday logic, when we look we influence events! Believe me this is the logical conclusion. It does not mean it is correct but it does mean that according to the logic we use it is the only answer we have so far come up with. This means the way we think is wrong, or the way we have thought so far is wrong and we need to find a new way of understanding.

The feeling among scientists is that there has to be something we have not understood yet, something which will explain it all. It defies logic. And so the search goes on to look deeper and think harder for it must just be hidden from view. What is certain is that there must be a meaning to all this, one we do not understand yet.

Scientists have found it hard to believe too and have continued to look for answers in what they call the unknown. Many have a belief that there must be something there we do not know, something we have not figured out yet. They call these unknown parts of the theory the 'hidden variables', implying they must be right there under our noses.

However, the idea that there were things we had not looked properly at yet were eventually ruled out by what is known as 'Bell's Theorem'. This was a touch of genius by John Stewart Bell in which he provided the possibility of a set of experiments which would prove the existence or non-existence of so called 'hidden variables'.

In his TV programme, The Secrets of Quantum Physics, December 15th 2014 Jim Al Khalili uses a game of cards as an analogy to explain Bells theorem. What it clearly establishes for me is that the theorem was first and foremost a matter of comprehension. The maths came later, as did the experiment. He eliminates the result of bias, or rigging, to determine his choice when dealt a hand and goes on to change the rules of the game by waiting until the hand has been dealt before deciding what hand to depict. He then points out that if he achieves a measure of success by a certain percentage this will prove Einstein's theory that the cards were preordained but if he loses by a similar amount it establishes that Einstein was wrong and that nothing in the entire universe was responsible for the outcome. It was pure chance! The latter having now been scientifically proven. You can see the experiment and the debate on Utube (5)

Bell was correct and it has now been established that there are no hidden variables – things we do not know about which would explain it all. So all we have left in our search for an explanation is to accept the strangeness of it all and look for the answer somewhere else. Einstein, did not like this conclusion and came out with the statement 'God does not play dice'. Neil's Bohr suggested he stop telling God what to do!

At this moment in time it looks like that the throw of the dice is fundamental to the way in which the universe is held together. That makes reality a very slippery concept. Neil's Bohr saw reality as just a word which we had to learn the right way was to use. Trying to get your head around 'Spooky Action at a Distance' makes his comment a serious contribution to the debate.

What I believe is of importance for a new understanding of reality is the acceptance, now proven in experiments, of the idea that particles are connected in some way we do not yet understand. The measurement on one particle can affect the measurement of the other particle. We know that they can actually look at one particle when it's perhaps miles away from the other one and find that the measurement of one affects the measurement of the other. We cannot avoid the conclusion that the two little chaps are in some way connected. In fact, they have a name for it and call it 'superluminal quantum connectivity', meaning faster than light. This is an amazing concept and you can look it up for yourself on the net. Its importance lies in the fact that it is not only the connectedness of the two particles but also that they can measure the state of the first twin and find that if they immediately measure the state of the second twin afterwards it shows that if there's any signal between the two that it must travel at perhaps even up to a million times the speed of light – that's putting a speed on spontaneous!

Hopefully you are now intrigued by all this and more prepared to consider the possibility that we can better understand coincidence by accepting the possibility that it is somehow related to the phenomena we know as 'spooky action at a distance.' (6) A fact which has been proven over and over again. That is the direction we are now going in.

Spooky Action is an earth shattering idea and takes quite a leap of the imagination, so in order to make it a plausible hypothesis we need to first off all reach an understanding of what is meant by a concept which is so mind blowing in its implications. To do that we need to enter further into the world of science and discuss some of its major issues, especially the nature of light waves

I will end this chapter by returning to my theme of coincidence. It is a term we use to manage events we do not understand. Its stops us looking further and leaves us satisfied there is no mystery attached too events, all is comprehensible. For example imagine a hypothetical scenario that may or may not present itself in the future. Imagine we have sent up astronauts to take a closer look a planet in the solar system. While there our astronauts encounter other life forms also looking at the same planet. We can at present safely predict what the headlines and the general consensus which follows will be 'Incredible Coincidence!' That is at present the only theory we have if such an event occurred. I am proposing another way of looking at what we call coincidence. In short it is possible a window into the real nature of reality and one which may at present hover only on the fringes of our consciousness.

CHAPTER FIVE
THE HIGGS BOSON

There is an enormous, complex and incomprehensible amount of information relating to the subatomic world and there is no way we are going to be able to digest it all. However, we can obtain a basic understanding of the more important and interesting findings currently being debated and argued over.

We are all aware that the universe appears to be made up of little round dots perceived by most of us as balls called atoms or particles. This has been the view for a long time, certainly since John Dalton the British physicist developed modern atomic theory over two hundred years ago. Over the last few hundred years it has been refined and added to until it is has now been promoted to what is known as the Standard Model. (1) This has led to the view that all forms of matter are made up of complex interactions between almost 200 elementary particles. This applies to us too. We are not quite like the goldfish but we are connected to the earth and its fundamental structure through the sub atomic world. Every breath you take is made up of molecules of oxygen and each molecule is made up of atoms and somewhere beyond the trillions in each breath lies the mystery.

Atoms are the ones we all think we know about and these are made up of electrons zipping around a nucleus made up of other chaps, some called quarks, and they are all bound together into larger particles called protons and neutrons. That's as much as you need to know, except that a proton is seen as having no mass and all these particles have anti-particles which doubles the number of particles in the theory.

The number and complexity of particles has led to them being called a particle zoo, a term which has evolved as scientist have been confronted with new challenging questions in their pursuit of the theory of everything. We have no need to delve too deeply into this particle zoo but we do need to acquaint ourselves with some of the theories and learn about some of the issues. We can do this by looking at the latest developments and the latest discovery, the Higgs Boson.

Having said that I am sure the interested and curious reader would like to learn more and would want to obtain at least a basic understanding of the standard model so that will be the purpose of the next few paragraphs. There are also many websites you can use to gather more detailed information about particle physics. (2) However, please bear in mind that will not need to learn any of this to follow the rest of the book but we do need to do justice to what is known as the Standard Model – a model I, and many others believe, will not survive the next century without radical change. Any details I provide are as much to try and educate the reader as they are to demonstrate the seriousness of, and the basis of, scientific theory.

Open any explanation of the particle world and you will read that the universe is believed to consist of four fundamental particles, organized by four fundamental forces, or messenger particles, which allow particles to interact with each other. This particle world is a buzzing thrashing wave of movement where things change all the time and particles can also change and modify each other. The question is how? The answer appears to be through other particles called force or messenger particles. These messenger particles are important to the standard model as they carry the messages between particles. If we didn't have them we would be left with the question, how on earth do particles communicate?

Gravitation, electromagnetism, strong nuclear, and weak nuclear are the four fundamental forces. Gravity, is the weakest of the four forces. Every object in the universe exerts a gravitational force on everything else. It keeps our feet on the ground and is a bit of a mystery and does not fit into the current theory that everything in the universe is made up of particles. I repeat, they do not know what gravity is! It is believed that gravity has a 'messenger' a sort of go-between in the exchange of matter which is called a graviton. This has not been found yet and so we just have to pretend it is there.

Electromagnetism, which also has a messenger particle, is the next strongest force and this is much different. Whereas gravity always attracts, electromagnetism has two charges: positive and negative, and it can either attract or repulse to the extent that forces cancel each other out, just like two ordinary magnets, pole to pole.

Eelectromagnetism is the chap they believe that keeps atoms together: the positively charged bits and the negatively charged bits are all attracting and repelling each other in such a way as they all keep their distance – reach an equilibrium.

So far things are fairly plausible but the next two forces are a bit more challenging and I mention them in order to demonstrate the truly incredible complications central to the standard model, they are the weak and strong nuclear force. The weak force works over sub-atomic distances and is found in an atomic nucleus and is responsible for radioactive decay. Its messenger particles are called bosons. The fourth is the strong nuclear force which like the weak force operates over subatomic distances. Its job is to bind things together in the nucleus of an atom and to keep another collection of characters working together called 'quarks' which operate inside protons and neutrons. The strong force has a messenger particle too. It is massless and called the 'gluon', which as a strong force holds elementary particles together. I must point out again that massless things can exist, especially as ideas.

These four forces dictate what happens in the subatomic world and the complexities of the atom and its numerous particles and the vast array of terms can drive you nuts if you are not scientists, but that should not deter the reader from trying to grasp an understanding of what is being said, although do bear in mind nobody knows how it all comes together so you are not alone.

Until reading about the subatomic world I always believed that particles inherited their mass at the beginning of the big bang. This is not now believed to be the case. They eventually came to realise that the big bang was limited in its potential to produce everything needed for the universe in one go. The scientists believe that differing temperatures over the first few seconds meant it did produce a lot of things like hydrogen and helium but the heavier particles operating in our universe only came along through the activity and formation of stars and galaxies later on.

It all sounds like they know what they are talking about and we are well on the way to discovering the big picture about reality. The amount of knowledge built up over our own lifetime in the world of physics has been incredible. This is true just in terms of what we know about the atom itself. Its nucleus is seen as a small, dense centre where nearly all of the mass is contained. The nucleus is thousands of times smaller than its radius - like a needle in a haystack – even smaller! In the atom there are two forces at work but the atom itself is made up of different types of matter - protons and neutrons, quarks and anti-quarks. Protons and neutrons exist in pairs, quarks and anti-quarks exist in groups of two or three. Quarks and anti-quarks counteract each other, meaning that every anti-quark contains certain properties that are the exact opposite of those in its corresponding quark. The combination of quark and anti-quark forms yet another type of subatomic particle which are very short-lived and so on and so on...

Most of what I have just written will be meaningless unless you are a physicist so do not waste your time trying to understand it. The detail of the sub-atomic world is best left alone unless you intend to take up particle physics. However, it is worth looking more closely at what is known as a Quark. These are very mysterious creatures which can never be seen. They have names like charm, strange, up and down, top and bottom. So far it has been impossible to get one to open up and sit still long enough. Most of what we know about quarks is deduced from the behaviour of other particles. In one TV programme they were likened to the Cheshire cat in Alice and Wonderland who disappeared in a puff but left behind his smile. The problem is that energy brought into a nucleus to try to separate quarks increases the force between them. At high enough energy, the addition of energy creates new particles so you never get to see the quarks as they can never be freed. I repeat nobody has ever seen a quark but they are confident that they are there.

Before continuing we need to get our heads around a few tricky ideas. Firstly, how it is possible to have no mass? Mass is anything that has weight and occupies a space. It can also, as Einstein pointed out, be converted to energy. Light has to be seen as both a wave and a particle, a bundle, packet, quantum of energy. It is always in motion and no matter which way you look at it always travels at the same speed. The law is that any object with mass must travel slower that the speed of light because momentum creates greater mass and so you just get heavier and heavier, infinitely at the speed of light and that is against the law. In order to travel at the speed of light you cannot have mass, nor can you be at rest. This means light photons are very much different to ordinary matter and as such they are free to behave differently and in ways we do not understand. Light is not alone as a massless particle as the gluon, the carrier of the strong force, is also thought to be massless. Do bear in mind that particles travelling at the speed of light can never be stopped or slowed down for observation as, we have to assume, they would immediately disappear. In fact they have to infer the presence of a single photon. When the photon hits the target, a metal plate, it causes a stream of electrons to provide a current and bring about a clicking sound on the plate. It is that sound which signals the presence of a photon.

I must also add that movement and activity of particles is in and between fields. Think of fields as like a vibrating drum skin. Sometimes a 'virtual particle' is emitted and absorbed by another particle. A virtual particle is like a ripple in the field, a bit like the hoot of an owl on a dark night. You cannot see it but you know something is there. The disturbance in the field tells you something passed through – especially if you can do the math's equations apparently.

Another way of understanding it is that whereas a particle is a noticeable ripple in a field just like a ripple on a pond, a virtual particle is different. It is a disturbance in a field that that is caused by the presence of other particles or other fields.

In other words a virtual particle is actually a different event from a ripple on a lake. It is something not understood and could be anything at all. The only thing they know is that it occurs around the same time as the ripple in the pond. That means it could be the cause of the ripple or be caused by the ripple or be the tail end of the ripple or a part of the ripple or be another current flowing past or through the pond. More graphically, the ripple causes another ripple, the ripple causes the fish to make another ripple, the fish causes the ripple, or the ripple is the effect of an unknown fish. Something else is in the pond?

Leaving aside the complexities of ripples and virtual particles we arrive at the really big question, where do all these particles get their mass from. They do not get there mass from other particles. Particle, X may turn into particle, Y but where did the original mass come from? It's a bit like a postman who keeps everything connected up by allowing letters and parcels to be passed around. The question is however, where the postman gets his parcels from. In this case it is from another postman – the messenger!

More scientifically, the answer is another particle. This other particle gives other particles mass by creating what is known as the Higgs field. Every particle in our universe gets its mass by submersing itself in the field. Not all particles pick up the same mass and there are a large number of differences from hardly any to thousands of times more depending on the particle. In addition some particles have no mass and are totally unaffected. The field has only one function and that is ensuring mass for different particles. The Higgs boson is also a particle and it has to get its mass in the same way as all the others by immersing itself in the field. This means of course that if we want evidence of the Higgs field we needed to find the Higgs Boson. This is made even trickier because once created it will decay almost immediately – nothings easy!

There are a couple of analogies for the Higgs Field, one being that you imagine some famous person walking through a room full of people. By the time they get to the door on the other side they have been sidled up to or bumped into a lot of people and picked up a lot more support.

The best analogy is by John Ellis a theoretical physicist at CERN. He asks that we imagine an infinite snowfield, the skier will just slide across the surface untroubled and unaffected by the snow, like a photon at the speed of light. A person with snow shoes will interact and pick up a layer of snow or mass, the person with boots on will gather a lot more snow or mass and be much heavier. In the analogy the boson is the snowflake and the snow is the Higgs field. (3)

The Higgs theory was proposed by British physicist Peter Higgs and teams in Belgium and the United

States in the 1960s in order to explain how particles gained mass. The extremely expensive atom-smashing experiments at CERN, the European Center for Nuclear Research, now claim to have captured a glimpse of what appears to be a Higgs-like particle. In other words they have identified all the bits and will now hopefully be able to see the big picture. I must inform you that new issues appear almost daily, so much so that the search for the Higgs was the easy bit. That is because in order to be the missing piece the Higgs has in some way to got to be responsible for creating matter – it's a bit like a Little Bang theory of the Big Bang.

They have certainly discovered something but David Charlton, among others, from the University of Birmingham who has worked on the Higgs at the LHC points out at the following site that there are still a number of problems before the issue is finally settled. (4). One being that the Higgs is far lighter than predicted and that they still cannot explain where masses much heavier than the Higgs comes from.

Recently they have put out a world-wide request at 'Higgshunters,org' consisting of thousands of images of collisions in the hope that the general public will spot something the computer missed. It is free to log on so try it out if only to see where the current evidence for most of the evidence comes from. Of course it is also understood that the Standard Model and the Higgs boson tells us nothing about a few concepts we have not discussed yet, Dark matter and Dark energy, the other 90% plus of the universe.

The Higgs then is seen as the missing piece: it gives the messenger or force particles mass and equips them for the job. It is now officially called the 'Higgs Field' and it appeared a trillionth of a second after the big bang caused by a quantum ripple.

As I have previously pointed out much of what I write about here is available on the TV or through a search on the net. I watched a TV show recently which reminded the audience of the fact that when the cosmos was only the size of a grain of sand, 'ripples' were created in the fabric of space. By a wonderful feat of their imagination they hit on the idea that the 'ripples' arrived in the first billionth second of time. I wonder sometimes if I am far too cynical?

While writing this I was able to listen to Peter Higgs on Radio 4 February the 18th 2014 being interviewed by Jim Al-Khalili in the Life Scientific series. If you listen to the interview, which I have included in the bibliography, you will learn the following:

"Like all elementary particles I think we field theorists think of it like Einstein thought of photons, these particles are just packages of energy of some kind of field and the feature which distinguish this kind of theory which leads to this kind of symmetry

breaking is the existence of what we theoretical physicists call the vacuum which means nowadays something different from what it used to mean. It means just the lowest energy state that you can possibly have in which there are no particles around but there may be something around and that something around can be a background field of some sort which pervades the universe.so in this theory there is such a background field and the background field in interaction with all the other stuff which goes through it is responsible for generating masses and mass differences of the other elementary particles the ones which are packages of energy of all the other fields. Simply because the background affects the way they waves propagate. But then the field itself can be excited, (or classically to view waves) to the packages of energy that are the Higgs boson. So it's an extra which comes with this type of theory you need to have something there which is the excitation of the background field." (5)

Peter thinks he was able to come up with the idea of the Higgs because he was a little outside the system and untouched by the 'folklore and prejudices' of his colleagues, He had, he said, 'a fresh outlook'. He was then asked to try and elaborate further so a non-scientist could understand but was unable. However, he did criticize a number of popular analogies by other scientists as they suggested a loss of energy and a slowing down of the particles. He could elaborate no further but he thought the explanation by John Ellis, mentioned above, was the best so far.

What on earth you may ask is 'broken symmetry?' Surely he could have explained that to the non-scientist? It has been pointed out before that the language of the high priests serves to maintain the bureaucracy, keep out the outsiders and avoid challenges to the status quo. I am not alone in criticizing the language of science. Alan Alada star of film and TV who played Hawkeye in MASH and also the co-chair of the 2009 World Science Festival in New York, has been running a campaign to teach scientists of all types to ditch jargon and to use simpler language and analogies.

In a recent News Night program Alan refers to a contest involving 22,000 students around the country searching for simple ways to explain such concepts as "What is colour?" or "What is time?" He then went on to recall,

"...watching a congressional hearing on climate change in which, he said, "a bunch of scientists were trying to teach congressmen about the science of climate change and the congressmen were trying to teach the scientists about politics. It was as if both sides were speaking alien languages." (6)

'Symmetry breaking.' is an important concept and it turns out that the phrase means almost unimaginably small fluctuations at a critical point decide what happens next in the sub-atomic world. For most of us symmetry is something we get when we look in the mirror or when we are describing twins. But symmetry is also a principle, a law of nature, like gravity, which tells us that something does not change its behavior whenever it feels like. We have learnt that these principles can be worked out and applied to many situations. We can use these principles to understand what is going on in order to predict the outcome. Scientists do the same with symmetries in the sub-atomic world. I should imagine that it was partly through using symmetries as a form of template that the standard model was built up.

We are not talking about the symmetries of objects but the symmetries of laws. A law of nature can be said to obey a certain symmetry if that law remains the same whichever way we look at it. We all know about basic laws that relate to the planets and govern motion and gravitation. I was brought up to believe that the universe was harmonious and symmetrical and orderly. Some would argue this makes Mathematics a true reflection of that underlying order. However, within the universe and within the sub-atomic world it now appears that symmetries can be ignored if nature so desires. The most notable one being at the onset of the big bang. At that point 13.8 billion years ago it was thought that equal amounts of matter and anti-matter were created. But they now understand that they ought to have annihilated each other. It didn't happen and that was because of a deviation or difference, of some sort, happening. This is an example of broken symmetry which fortunately for us ensured our universe survived. Coincidently it was also rather fortunate.

Scientist now see non-symmetry as just as important as symmetry when trying to understand and explain what is happening in the universe. Things may have a tipping point whether it be standing on one leg or a boiling kettle. Things may have a knock on effect, a butterfly effect, or you didn't see the symmetry of this-being-broken effect. The idea of a harmonious, uncomplicated symmetry being the underlying structure of all things is wrong. Broken symmetry is necessary if we are to explain difference and uncertainty, unpredictability and separateness. This is just as true in the world of particles as it is in the universe as a whole. For example quarks, depending on whether they are positively or negatively charged do not always behave as they should when their actions are reflected in a mirror. They violate mirror symmetry! Imagine looking in a mirror and seeing yourself upside down. Spooky!

How these symmetries came about was once seen as a mystery until scientists decided that in many cases it was just an accident of nature and not a fundamental principle of nature – a bit like the duck-billed platypus, or if you are really cynical, 'because!' We may look in a mirror and see a perfect symmetrical image. This is not always the case in the particle world but instead of beating your brains out looking for the reason for this we just have to accept that some things are just an accident, or coincidence, and have nothing to do with the symmetry of nature. Most of the universe is symmetrical but this too may also be an accident.

Secondly, it is also worth knowing that empty space is actually a medium in which particles interact and obtain mass. Energy conservation is a fixed idea in which we assume it must be passed on and cannot be lost but this does not apply to the quantum world as, according to the theory, it can appear and disappear whenever it likes. That means that there is no such thing as 'nothing' which also means that a 'vacuum' is the lowest possible level of the universe. Peter Higgs recognized that in order to account for mass something else had to be there to bring it about and that is how he arrived at the Higgs Field. This idea of the vacuum is important. It is a recognition that there is no such thing as empty space. One way of comprehending this idea is to ask how gravity would operate across nothing. Empty space is now the vacuum.

The Higgs theory needed to be backed up with evidence and consequently the atomic colliders at CERN (the Center for Nuclear Research) which was very important in revealing the vast array of particles and their complex interactions, set about looking for the proof. They been working overtime and millions have been spent looking for this little elusive chappie - without it the Standard Model was dead in the water. Let me remind you, the theory is that the Higgs Boson, creates, enables, provides, and via a field which imparts mass to other particles as they pass through the field it produces. There was nowhere else to go. It was the logical conclusion of the way in which the present picture of reality had been built up over the history of science. The wonderful news is that just when they were almost ready to give up the search, evidence that the Higgs Boson really did exist was revealed to the world on the 4th of July 2012.

Further insight into the latest theories is possible by looking briefly at how they went about searching for the Higgs,

Many of these experiments were done in liquid hydrogen or liquid bubble chamber. It is heated up to the point it is starting to boil just like water does and that means it is predisposed to form bubbles about any changes be it bits of dirt or movement of any kind. Any disturbance will get a bubble around it so if you fire a particle through it tears up or knocks out other particles and it leaves a trial or track. We can then see the bubbles form around the particles which are left behind and take a picture of the event. But if the particle hits the nucleus of an atom you get a different picture and a formation which can look like a simple Y branch or a more complicated branch with many offshoots revealing other particles or disintegration of particles. This happens close to the speed of light in billionths of seconds and has led to hundreds of new particles being observed. These have been organised into patterns based on spin or mass which has led to new theories. For example in order to make sense of things scientist have had to produce the notion of 'quarks'.

So at CERN the photons from millions of atoms are sent speeding at almost the speed of light and crunched together, smashed into each other. At the point of contact they reveal information, sometimes based on colour or trajectory, which helps scientists build up picture which will confirm or negate their theories. You hit a single photon with another particle and you generate hundreds of different particles. Atoms once thought to be the smallest particles are quite big compared to these collider emissions.

All this information is available through Professor Micheo Kaku. He regularly appears on lecture tours and TV programmes and is always ready to help people understand what is going on in science and you can contact him through his own website. (7)

The findings at CERN revealed a wide differences in particle size and mass and so the zoo was broken down into families in order to make sense of it all. Some were given new names like up/down. Bottom/top, strange and charm. The family theory became a hit, especially after predicting and finding the sixth one called 'top', which took them about four years and was only discovered in 1990 with the aid of The Tevatron, the second most powerful particle accelerator, now shut down.

This particle which is seen as heavier than the other five and because it turned up in the first trillionth of a trillionth of a second after the big bang was difficult to find. Consequently, in order to see some evidence of its existence, they had to create billions and billions of collisions to see just a few of them.

The Higgs Boson is important for modern science as it's seen as proving that there is an invisible energy field that fills the universe. Nobody will ever see this field, the Higgs particle which remains after the particle collision is just an indication of its existence. The belief is that just after the bang, a trillionth of time after, the Higgs field switched itself on. What happened then was that some particles, quarks and electrons, note not all, began to feel like they were in a field as if in a swimming pool. The field, in short, held on to them and gave them mass. In providing mass it allowed them to come together and make up most of the atoms and molecules we have today.

So the answer to the question where do particles get there energy from is another particle which we have now discovered creates a field and as these other particles, which were massless, pass through it they pick up the energy. The Higgs Particle was created at the time of the Big Bang/Inflation and there were possibly other key field making particles created then which they are hoping to discover now that CERN has started up again. You can keep up to date with events at CERN through numerous websites. I use Matt Strassler's website. (8)

It was some time later in the history of the universe that other particles came along. However, it must be noted that the Higgs field making the universe what it is today started off within a trillionth of a second of the beginning. Without this important field building event all fundamental particles would weigh nothing and hurtle around at the speed of light and as such would rule out life as we know it. It would spell disaster for the formation of atoms, everything would behave like light, which science informs us has no mass. Also the theory itself would have to go in the dustbin of history. Thank God they found the Higgs!

However, the field does not affect all sub atomic particles the same way. Particles of light or photons, as mentioned earlier, feel no drag at all and remain without mass. Some scientist argue that assuming they have now found the Higgs boson they will still not be able to explain why some have more mass than others, which means that if they follow the same logic which got them to this point they may have to assume there are other Higgs bosons' they don't yet know about which were created at the same time as the Higgs, probably within a few trillion seconds here or there.

We certainly know an awful lot more than we used to. This wealth of knowledge has all come together in the last few hundred years. However, I must remind the reader that there are no theories that as yet have stood the test of time, so very recent ones are almost certainly going to be heavily amended or proved wrong sometime in the future. This leave the latest revelations about the Higgs Boson interesting, although you may think they are likely to be another example of the human race still possibly looking in the wrong direction, or even worse not yet capable of coming up with a real glimpse of the mystery of reality.

Having an invisible field which makes everything possible sounds very speculative but nonetheless the sort of thinking a goldfish would need to start his research. How did we get to this point in our thinking you may ask? A point where we depend on an invisible field to impart mass to the subatomic world, and only some of it. In fact photons are left to carry on moving at the speed of light forever, unless they fall into a black hole. It could be seen as making Alice in Wonderland a fairly lightweight use of our imagination, but I am keeping an open mind.

There are a number of questions by scientists which as yet nobody can answer. These questions make the theory either very much unfinished or very unsatisfactory. These are complicated and to do with the mass or non-mass of key particles, including ones like the graviton which has not been discovered yet. What puzzles me is that the 'field' came about somewhere at the beginning of time so we now don't just have to try and imagine something arising out of nothing but now its two things arising out of nothing, the universe and the Higgs field, more or less at the same time – bearing in mind 'nothing' is a vacuum with lots going on inside it.

There are also a number of other questions which come to mind. Why are there two forces in the nucleus which balance each other out - one repelling and one attracting? That's a very fortunate, actually amazing, coincidence. And how absolutely mind boggling that the slightest of changes in the balances of forces and none of us, not even the universe, would exist. It the same with gravity if it was any weaker or any stronger than it is that too would finish us all off. Everything appears to be so complicated, so dependent on images of fields, mathematical equations and faith that some things exist even though we cannot see them. There is lots of room for believing we may be on the wrong track.

By now you realize that according to the standard model we should be thinking about fields as every particle has its own field and every particle has to pass through the biggest field of all in order to obtain mass and that includes the particle which is responsible for the most important field of all, the Higgs boson. However, the media today is talking about a couple of new Noble prizes, one to Takaai Kajita and one to Arthur McDonald for discovering that a particle, the neutrino, once thought to be massless does actually have a mass which is a million times less than the electron.. They pass thorough us in their thousands every second and are extremely difficult to detect. It appears to be a finding which messes up the theory of the Higgs field a little – unless they find a new field. (9)

The standard model is not the only theory on the block. In fact there almost as many theories as there are physicists and some of them, which we will be looking at in the next chapter are, to use a favourite term of Richard Feynman's, 'absolutely mind boggling.'

CHAPTER SIX

OTHER ISSUES AND OTHER THEORIES

It is now, in this generation and thanks to the media and especially the World Wide Web, that we all have access to the latest experiments and ideas.

Last night I watched a TV programme which was all about the ways in which the media had brought about this change in our lives and had made it possible for us all to be involved, if only as an observer of the massive growth in science and its many and various disciplines.

So far there is no grand theory. At the moment The Standard Model is the basis for most research as the conviction is that once they find the missing pieces they will be able to sit back and rejoice in the knowledge that they now know what it is all about. Unfortunately, as I have pointed out, the evidence so far is that they still have a long way to go. In fact some aspects of the theory can never be proved and rely totally on a hypotheses and that may always be the case.

That is problematic enough but the really big problem for The Standard Model has been to explain where mass comes from. Without an answer the Standard Model would have collapsed and that is why it is so important and why they have been searching for it for the last fifty years. In the past they have tried to resolve the issue through mathematical equations, but things have not really made sense However, after scouring through the debris of millions of tests in the CERN atom collider they believe they have possibly found the answer. Further experiments will have to be done to confirm it but what they are certain about is that they have found something and, as explained in the previous chapter, it is called the Higgs Boson

Before moving on lets briefly look again at what was said in the last chapter. The Higgs field is essential to the belief that there was an event. The theory holds that just after the bang, a trillionth of time after the 'pause', the Higgs field switched itself on. What happened then was that some particles, quarks and electrons, note not all of them, began to feel like they were getting slowed down in the field as if in a swimming pool. The field in short held on to them and gave them mass –which we all know is to do with energy. This slowed them down and in so doing allowed them to come together and make up all the atoms and molecules we have today. Without the field all fundamental particles would weigh nothing and hurtle around at the speed of light and as such would rule out life as we know it as it would spell disaster for the formation of atoms. Interestingly, I remind you, things going at the speed of light have no mass but if you were able to stop them they would disappear – now there's a twist in the tale.

However, the field does not affect all sub atomic particles the same way as some, such as particles of light or photons, feel no drag at all and remain without mass. Without this field we would not be here. In order to add to your confusion you need to know that even if the Higgs boson is confirmed through a lot more experiments they still have no theory that will explain why some particles have more mass than others. The Higgs is the most vital part of the Standard Model and a few unanswered questions will have little influence on the debate. In fact they are already talking about other 'Higgs Bosons' they don't yet know about.

By now the reader should be aware that the search for the grand theory of everything has a long way to go. The Standard Model is not complete and scientists have identified a number of problems we need to remind ourselves of. It cannot explain gravity and where it fits in, it cannot explain Dark Matter and Dark energy, which we will talk about later, and there are too many things we just have to accept, such as why the masses of particles are so different and 'the weak force comes out as ten quadrillion times as powerful as gravity for example'. (1)

We can also note that they may have found the Higgs Boson but there are lot more experiments to be had before they can confirm it. They may yet find it is just another particle and there are many more particles to come – time will tell.

Not everybody is convinced that the picture so far, based on the Standard Model, is the full picture or the right direction we should be going in. Their doubts have brought about a consensus amongst the best brains on the planet that we need to keep an open mind about what the future may hold, but more importantly a recognition that we cannot rule out a far more incredible and complex underlying reality. With that in mind we need to look at a theory which many see as giving us another direction to look in, String Theory.

The strings are thought to be one dimensional, just a length without height or breath. There are two types, closed and open, one a loop and one open ended. To understand how it works we have to forget about particles as points or things and think of these strings operating below the upper levels of the atoms containing atomic nuclei. We need to get down below the very smallest of the inner parts, to the vacuum. The string is compared to a musical string which vibrates. These vibrations or musical notes are what we normally assume particles to be, but in string theory what we refer to as particles are in fact a result of the behaviour of elementary strings. (2)

More scientifically, particles in string theory arise as 'excitations' of the string, and included in the excitations of a string in string theory is another type of particle: this particle carries the gravitational force and is called the **graviton**. In the standard model the mathematics for the graviton don't work but in string theory they do and that means they can include gravity in the grand plan and argue that string excitation carries the gravitational force. Brian Greene on a U-tube video points out String theory came about when scientist realised that any unifying theory needed to explain where gravity was going to fit in. Einstein had established that gravity was caused by the curvature of space and objects rolled along their local curve or valley. The theory was proven correct but the question remained, how did it fit in to a grand theory of everything? The answer was deemed to be through mathematical equations, but to make it work they had to suppose there were other dimensions. Brian Greene points out that theorists found an equation that worked and then noted that it also fitted beautifully with a description of the electromagnetic force, the force explaining light, radio waves and electricity in general. This was looked on positively by the world of science and the theory became much respected and was treated as a real contender for the theory of everything.

There are a number of excellent videos on Utube and I recommend the following one where you can watch an interview between Brian Greene and Lawrence Krauss. (3)

He explains that within the basic constituents of matter, the sub atomic particles which are almost like dots, there are tiny strings which are so small they may never be seen. These little strings can vibrate and the result is one of the many particles which have popped out of the particle zoo. The theory has been developed further because in order to make the math's work it needed other dimensions where strings might also vibrate. They started out with 22 dimension, 26 if you added time. This has now been whittled down to 10. He likens it more to a hypothesis than a theory and it has now moved on to suggest new dimensional shapes like Branes or sheets. The Branes may be big and support our universe. It may be just like a slice of bread, we live on one but we cannot see the other because the photons in our slice only works for us. Possibly one end is attached to us and things cannot cross the divide.

Micho Kaku points out that Einstein's theories break down at the Big Bang, and with black holes, and this was another factor paving the way for String Theory. Check out his site for another good explanation of String Theory. (4)

String theory describes what happened before the big bang and suggests that we are just one universe alongside other universes. This is a compelling theory with massive implications. We can measure a house, its width, depth and height and all its dimensions including every object inside it. We can then change and alter these dimension to get any shape we want while all the time retaining the concept of a house. They are unable to do this with a model of the universe as when they try to rearrange the universe they find that it is such a fine balance that if they change it in any way at all they cannot create a universe which will sustain life. In other words we are obliterated from the scene. As a way around this many theorists have come up with the idea that there are many different universes. We happen to be the one with life on it. There may be others and if that is correct then it follows logically that there is life in some of these other universes. The problem is that there is no evidence for them but as there now appears to be a world below the subterranean depths of the atom and the mysteries of the quantum world the theory is a real runner.

The reason they cannot rearrange the universe relates to what is known as the 'fine-structure-constant'. So we have to entertain a little bit of math's here. This is seen as the most important of one of many puzzles. It has the number 137 and this governs how particles interact. It has also showed up in many other measurements like the mass of quirks and electrons, the strength of gravity and the electrons magnetic force - all of which are essential when trying to understand things. They know the values of these measurements but do not understand why this value and not another. Feynman calls it 'the greatest damn mystery of physics'. It is a constant number although some now think it may have varied over time as the universe evolved. By now the reader hopefully has learnt that theorics always appear to be in a state of flux. It is also a dimensionless number as it has no units and this suggests it possibly is a bit like 'pi' which we can use to find the diameter of any circle. Some think that 137 may have a similar function and may be a key to unlocking more secrets of the universe.

Currently nobody knows where this number comes from but its implications are enormous as if these values were slightly different we would not exist and neither would our universe. Brian Greene provides an analogy with the issue when he tells a story about his very young son asks how come the store had his particular size of shoes. We have the answer to that but why the magic number in physics is 137 is a mystery waiting to be solved and the only answer we currently have is it is either a fluke or a coincidence. Of course in the multi-world theory, we will talk about later, 137 would just be one of many possibilities. (5)

String theory was a way of bringing all four forces together and at the same time to include gravity in the explanation. The theory suggests that the Big Bang came about when the universe collided or separated, from another universe. There is more than one universe and some of them are very, very little and may even be curled up, which in terms of relativity would not matter.

Over a very short period of time the theory was built on and added to include what is known as Super-symmetry. This predicts that there are two main particles which are connected in superstring or super-symmetry. These were discovered through mathematics and nobody has ever seen them yet. The belief is that for every particle which transmits a force there is another one which makes up matter and Super-symmetry is the connection between the two.

Initially the theory was all about particles like photons which are massless but there was no explanation of how it related to particles which had mass called fermions. These fermions are the building blocks of atoms or matter. It's a bit like how do closed strings become open and open strings become closed - what kicks it all off? They use the word 'perturbation' which sounds like a form of anxiety. They had hoped that by now atom smashing at CERN would have revealed something to support the theory. The proof would be evidence of the graviton. The graviton can move across dimensions called Branes and they are hoping that some evidence from CERN will indicate they are on the right track. So far there is no evidence and it all hangs on the up-graded Large Hadron Collider and the next batch of experiments just started.

Super-symmetry is under the microscope at the moment. Peter Higgs mentioned in his interview that they had hopes for the theory of 'super-symmetry', the idea that every particle has a super partner with a higher mass. If found it was supposed to resolve a number of problems in the standard model where the Maths did not work out. Unfortunately, the mass of evidence accumulated in the search for the Higgs has so far thrown up nothing indicating the theory of super-symmetry has a future. The following quote is from Professor Jordan Nash of Imperial College

London, who is working on one of the LHC's experiments,

"The fact that we haven't seen any evidence of it (supersymmetry) tells us that either our understanding of it is incomplete, or it's a little different to what we thought - or maybe it doesn't exist at all," he said.(6)

However, I must point out that many physicists believe that the reason no one has observed the particles yet is because it takes a lot of energy to generate them.

Despite the lack of evidence for Super-symmetry String theory has not gone away and now new ideas have made it an even more interesting and thought provoking scientific theory. The latest development we can look it involves 'Branes' which I mentioned earlier. These are now seen as an essential part of string theory, now called M theory. (7)

There have been many types of string theory over the last forty odd years but they eventually settled on one, mainly popular because it was the only one which allowed for quantum gravity. This M theory allowed the strings theory to be attached to something called a Brane – like a membrane or two dimensional big sheet of paper. The other universes were likely to be attached to the end of the Brane. Many equations did not work, did not make sense until they allowed for ten Branes which are really 10 dimensions of space and one of time. To exist at all, according to the equations, they have to be a millionth of a billionth of a billionth of a centimetre. That's going to be hard to prove!

String theory, upgraded to M theory is a very imaginative idea and it presents us with the possibility that things do not move only within dimensions but also between and possibly across dimensions. This would suggest things or events are able to cross paths on the journey through time. This means there is possibly movement not just forward within a dimension but across into other dimensions. Another way of understanding what this gateway of an idea means is that if things occurred in only one dimension they would be there transfixed forever. It is only by having a multi-dimensional universe that events are able to keep moving and that time, as we see it, constrains us to see it moving in only one direction.

We need to keep in mind that we are talking about theories, but as you can see the Standard Model is not the only theory in town. There are a number of alternative theories about how it all began, many of them from String Theorists and its various branches, but all of them are by supported by different highly respected professors of physics. In fact it is now quite common among the theorist to argue there was something in place from the start, before the so called Big bang. In many ways this thinking has developed since Hubble pointed out through his exploration of what was happening to the universe and noted that the universe was expanding. If expanding it must have had a beginning. However, as this makes it all effect and no cause it eventually brought about the realisation that there must have been something there before it all began.

The following theories where all picked up while watching an American TV series called Through the Wormhole by Morgan Freeman 2010. (8) One theory by Professor Andrew Lyndsey argues that we should seriously entertain the idea of 'eternal inflation'. The universe is like a bubble in a big block of cheese, and just as the cheese has many bubbles then there are many universes. Others disagree and one argues for a big bounce instead of a big bang – meaning that the universe is forever bouncing back and forth. Professor Lee Smolling believes general relativity is not a complete theory and that the universe reproduces itself through black holes, again and again and again.

Professor Neil Turok of Cambridge University cannot accept the big bang argument that there was no time before the beginning. He says,

"There have been many Big bangs and to believe there has only been one is an illogical contradiction. We have to ask the question, why these laws and not others."

The professor, an advocate of String Theory and Branes, argues that this requires that the universe came out of a pre-existing universe and, he says, *'this requires ten spatial dimensions, plus time.'* We live on an extended object he calls a membrane. This has three dimensions and that's part universe as we know it. There are two of these membranes separated by a narrow space which is another dimension. When these membranes collide they create something which can then fill up with a finite density.

The theory of Laura Mersini-Houghton, a professor at the University of North Carolina is made much more important because she claims it is backed up by evidence. She argues for a multiverse. Her theory demonstrates the presence of neighboring universes which she argues can all be explained in detail, and these universes have a gravitational influence on our universe. It is an idea made more feasible by its ability to answer three observations which have so far defied explanation. One is a void, or patch of nothing, in the cosmic background, second is great swathes of galaxies where six have been found to be doing their own thing and moving in the wrong direction, and thirdly there is something odd about the temperature in outer space. These events are all due to the behavior of neighboring galaxies and can be explained in precise detail through her theory, she argues.

Professor Sir Roger Penrose was once very scathing towards people who were unable to accept that it all started with a big bang. He had argued the case most of his lifetime. He believed that a body collapsing under gravitational pull will end up as a singularity (disappear). However, one day he had a think about how the universe might end and that helped him to change his mind completely. The theory he once loved was that as the universe cooled down all that would be left would be mass-less photons. Without mass there would be no time and nothing to measure, no dimensions, and no idea of size. He now argues that our universe is but *'one ion of a succession of ions'*. Meaning this universe will become the big bang of the next. All the mass of this universe will be converted to energy, the starting point for the next universe.

These different ideas legitimise my own speculations that, despite the lack of research findings, that the future lies in a new and challenging idea – that may or may not be String Theory. My belief is based on the notion that the Standard Model has reached the end of the road, there are a number of unresolved questions I have referred to which mean it is unable to go any further without massive and radical changes. My other reason for looking elsewhere is that nothing stays the same and no theory has yet stood the test of time, we are after all, on the first pages of history.

I must emphasise again that String theory is a theoretical model which makes mind boggling claims and as yet there is no evidence that it is the way forward. I remind the reader that it suggests that there is an underlying reality which can be compared to a piece of string so minute, curled up or open ended, you cannot ever hope to see it. It is almost impossible for us to imagine it exists without at least one of the four aspects of our dimension, length, breath and height. Also bear in mind that not one of our three known dimensions can exist without time.

If future research lies in string theory then we are talking about an area of activity, the vacuum, which always is and always was. There is a lot of activity in this subterranean world below the particle world and it too is part of our dimension, if only because any activity, including waves, needs a dimension to move in – that is the vacuum. However, String Theory has far more to say about dimensions. In the universe we know each object has a dimension of its own within a dimension, but what is being said here is that there are possibly dimensions outside of ours running parallel to our own. This is not an isolated view but one that is seen as a very real possibility. We can only speculate about the size of these dimensions, but I repeat, they theorise 10 and one of time.

Within this very broad area of evolving and competing theories there does appear to be one thing that all they all agree on, whether it's the standard model or string theory, which is that below the world of molecules, atomic nuclei, and quarks there are fields or vacuums we know very little about. But we do know that they all participate in an incredible and complicated, active flowing, always moving, web of information. We also know that, within this world of waves and particles, if two photons or any other particles are separated, even if a light year apart, a change in one creates a simultaneous change in the other. That is a wow!

I believe that these ideas make possible the notion that what we call coincidence is a key to greater understanding. The odd word or idea you hear again and again on the same day, the person from next door you meet on the other side of the world, the sheer unlikelihood of two similar events which we have all experienced; this means there is possibly another explanation we are as yet unaware of for the label, coincidence, but we will talk more of that later.

All these ideas and developments in science have happened over the last hundred odd years and while many experiments and theories have opened up our understanding the more significant of these for me has been the two slit experiment we talked about in chapter three. There are lots of theories about what happens in the two slit experiment and as yet there is not an answer which is universally accepted as being close to the truth. In fact theories are split along two lines, one plumping for staying in the world of classical science and ridiculing attempts to reach into what is seen as nonsense and the other view which, like me, accepts we have not even begun to touch on the nature of reality and we need to embrace the fact that there is a mystery we may, or may not, never fully understand.

It is now clear to us all that the world of sub atomic particles moves as waves. In short a particle moving as a wave can then become a particle and then move again as a wave. The theory is that when we look we choose for the wave a dimension it can sit in, or put the same thing another way, we collapse the wave and force it into taking sides. It is no good looking for an example which may illuminate our understanding in what we call the natural world. There is not going to be a moment of clarity where we say we have seen this before in some other aspect of nature. We have moved beyond that and into a realm where there are no examples in our experience of reality. Up till now we have just have to accept that by looking we have set the wave a task and it duly responds by sticking to our reality. Fortunately for creativity, not everybody agrees as you shall see in the next paragraph.

David Deustch in his book The Fabric of Reality, differs from the common, so called Copenhagen Interpretation of the two slit experiment, that when we look we determine things, and states in his book quite scathingly that the problem was that;

"...its motivation was essentially to avoid the conclusion reality is multi-valued and for that reason alone it is incompatible with any genuine explanation of quantum phenomena." (9)

David has a different theory and makes a very clear argument for parallel universes – the multiverse. He spends nearly all of Chapter Two to establish through a version of the two slit experiment that logically and conclusively there are other particles outside our universe which are responsible for the strange behaviour of light. This applies not only to photons but to all particles. Bit by bit he demolishes all the arguments which may undermine his assertions. Finally for the purpose of his argument he invents shadow and tangible photons and demonstrates through his paradigm how they are both tangible in their own universes.

David Deutsch also reminds us that quantum theory as presented through the Copenhagen theory, *"applies only to unobserved aspects of physical reality"*. In other words the idea that which way the ball bounces depends on human observation is all they could come up with. David Deutsch gives us a choice between believing in another parallel universe, or universes, or in believing we change the order of things when we look at them. David writes about 10,000 words on his observations and you can read it all in his book. In short he is arguing that there must be something there responsible for the strange goings on and this can be explained by recognising it can only be occurring in another universe.

These ideas are difficult to accept not just because they appear as mad or incredible but also because we are confronted with a socialisation process from the moment we are confronted with life. This ensures we are all keyed in to the need not to avoid making waves, to keep in line and not rock the boat. The prevailing ideology cannot be underestimated as it is an enduring part of the human condition which inevitably, at certain times in history, will be a factor in holding up progress, and sending us in the wrong direction by ignoring new ideas and new developments.

There are other issues with our logic we have to question if we want to ask questions, the way we have been taught or learnt to reason that things should behave in a particular way. For example the best analogy we have for the behaviour of light is with waves in the ocean, but it is not quite right and so, as the two slit experiment shows, it defies our logic. The point is we now have to use a different research tool as the old one no longer works – we have moved into new territory as all this implies there is another reality, operating out there somewhere. Some, like the American Hugh Everett, who try to answer the question of what happens to history, would also suggest that something that was in two places at once gave some credibility to the theory of a multiverse - an unimaginable amount of alternative realities. This stems from the idea that a particle can be not only in two different locations at once but also have different velocities. Surely that involves another dimension at least.

Great, I am clearly not alone in saying we have to move on from a logic which no longer works, a logic programmed just like it is for the bees, the birds and the cuckoo. How would the cuckoo escape its condition? It cannot! Yet we suppose that we are free to know everything. Clearly we need to think outside the box, to think very differently, reach out beyond even what we can imagine now and look for clues everywhere and anywhere.

We know that we interpret the world through our heads and so must start by conceding that logic is a human construct and does not and will not tell us everything we want to know about reality. It follows that it is already decided what we see and how we interpret with the tool we call head - unless we challenge ourselves the programme is set. At certain periods in history in order to get an insight into the world any other way is not possible through conventional means of reason and logic. In fact without reaching out into the extremities of our imagination I cannot think of any other way of seeing reality more or less the way it is now. It's the same for the goldfish – he has no options but to see it as the way it was set in his smaller brain. The question is do we see reality as it really is and all we have to do is piece together what we see before our eyes and one day we will understand it all. The answer could be yes but that would leave out what we know to be an option as the other answer is no. All our attributes, not just experience and logic count in the search for the truth. We must use every means at our disposal to unearth the soul of the universe. In the process we must also remember, we humans can make mistakes, misinterpret and misinform, and of course read into things exactly what we want them to be.

Limited as we are the only way to see things differently is through every potentiality of our imagination and even that may not be enough. It also follows that our imagination has to be stimulated by whatever reality throws at us. Of special importance is the stuff that totally baffles us. That is the best evidence we have: that we have not figured it out yet. If we believe that only what goes on in a laboratory or the minds of a group of bureaucratic scientists is the only way forward then we are doomed to mouth the same noise as the goldfish.

If there are other dimensions we do not have to suppose they are undetectable but we do have to suppose they are interdependent or at least have an interface. Remember there is no such thing as nothing. If another dimension had no interface, was totally independent of ours, then according to the world of science, it cannot be separated by nothing. That can only mean that there has to be an interface between us and the existence of another universe, otherwise we have only one universe.

The following Matt Strassler quote, stimulated me and set up an idea I find speculative but satisfying. He states the following.

"One of the great contributions to science of Nima Arkani-Hamed, Savas Dimopoulos and Gia Dvali was to observe that no one had ever excluded the possibility that we, and all the particles from which we're made, can move around freely in three spatial dimensions, but are stuck (as it were) as though to the corner edge of a thin rod --- a rod as much as one millimeter wide, into which only gravitational fields (but not, for example, electric fields or magnetic fields) may penetrate. Moreover, they emphasized that the presence of these extra dimensions might explain why gravity is so much weaker than the other known forces."(10)

What interests me is the phrase, 'only gravitational fields may penetrate'. I think there is another way to develop this idea. Let us agree that we are to a large extent at this moment in time constrained by our human condition and the circumstances of our universe. That does not necessarily mean everything is, especially consciousness. It is also possible that what we call the underlying reality is the home of the gravitational force. In other words the main, or major force which keeps us altogether, what we call gravity operates across every level of the vacuum identified by both the Standard Model and String Theory. That means bridges across and infuses all dimensions of reality.

If there are other dimensions it is reasonable to assume that other dimensions cannot exist totally independently of other each other. Something must lay down the law for them all, and some of these laws or at least one of these laws is possibly through the influence of gravity. Gravity may operate across all dimensions. Whereas light brings warmth and energy, gravity brings order. Photons, electro-magnetic waves, weak or strong force do not bother gravity: but gravity bends light and space. That makes gravity a very, very important player. If we do not know what it is it may be because it is not only of this universe but is the master of them all.

Such a theory can be seen as nonsense. However, apart from the enjoyment of exercising my imagination I must also point out that it is not really going much further than many of the widely different theories available. It suggests that there is at least one and possibly many other dimensions next door and parallel, and invisible.

If, as speculated, gravity has such a special role that would suggest that the idea of 'Branes' is something we should take more seriously. It is worth pointing out that this theory would not contradict or undermine Einstein's theory of relativity and the way in which space itself is affected by gravity as it would still affect space. Maybe we should not be looking for gravitons but ideas about how we may discover gravities management of the subatomic world.

Such research would of course be very difficult as if it is not an actual particle that is shaping and managing the sub atomic world then it is very unlikely it is going to be picked up by the Hadron atom smasher at CERN. This brings me to another observation, and an important one for the arguments being made in this book. So far the search for the theory of everything rests on finding the last particle, the one that fits it all together – hence its name as the God particle. However, it is very much a circular argument as they keep trying to explain the relationship between particles like it's an orchestra - the maestro being the Higgs.

The Higgs was supposed to be the giver of mass, created in a few trillionth of a second after the Big Bang. It now appears from very recent research that they will soon be starting to look for new particles to marry up the need of other newly discovered particles. That is the news from CERN and many respected physicist. The good news is they know how to look for particles but the bad news is they do not know how to look for anything else. You can check any of this information out by asking your search engine for the latest news from CERN.

Up to here

CHAPTER SEVEN.
THE QUANTUM BRAIN.

Imagine all the space in the universe with no thought. What on earth is thought, never mind atoms, what is thought? We may also want to ask what powers are these we have that can bring things into being when we look.

I have dithered about where to put this chapter as we still have much to talk about before we start on our own theories. However, with all these strange and incredible theories flying about it seems a timely place to talk about the human brain.

Without the human brain we are like a stone. It's the real wonder of the universe as it not only provides us with a tool for looking at the world out there, but also the world in there, the one inside our heads. Through the brain then we interpret the world and it is the brain which has the job of providing any understanding. It is difficult to think we can think outside of the brain, that we can reach a mode of thought which is unconstrained, unlimited by its condition, its parameters, built in paradigms, or predetermined constraints.

Once you put your thinking cap on you are left in no doubt that the only reality we can be sure of is that the brain is a network we not only see the world through but a network which influences the way in which we see the world. This is an age old argument and is best understood when we look for answers to concepts like time, space, cause and effect. These concepts are believed to be built into our understanding, they are tools we see reality through and like our ancestors we must take them as given facts of life. Each of these concepts is a mystery understood only in relation to something else, time without a beginning, a beginning without time; space without matter or matter without space; cause without effect or an effect without a cause, trying to understand one without the other causes a massive headache.

Taking our thoughts and ideas for granted is part of the human condition and raises the question 'what is thinking' and 'what is happening when we are thinking about 'thinking' a real challenge. It also means that we would be very silly to assume there are simple answers. Rather we should accept that at some stage in our thinking we will need to try and think 'outside-the-box' if we are to get closer to the real nature of reality. Maybe it matters not that we can, through metaphor and analogy, make use of our imagination by putting ourselves in another reality. However, I believe that to have got this far in the pursuit of knowledge and to be able to write this article so as to reach out into the unknown is evidence itself of a potential in something else we have no real understanding of, consciousness, and the workings of the human brain.

For example let us compare ourselves to the goldfish. Assuming he is fed daily at a given time and one day his food is stopped the goldfish will not get the right answer by some form of logic. Only a leap into the unknown may help him reach an understanding, and he may need many more leaps to get even close to what we know and what we believe is the truth.

It is our advantage as human beings that we can put ourselves into the head of the Goldfish and imagine the world through his eyes. In so doing we can imagine the leaps in consciousness he would have to make to obtain any understanding of the reality we are currently trying to understand ourselves. The thing is we already have theories of how it may, or may not, have happened: the Big Bang/Inflation, Evolution, Quantum theory… If we throw away our conceit and see ourselves in a similar situation to the goldfish as having to make huge leaps in our imagination in order to find answers to our questions then we then we may find some answers.

This is also the view of John Archibald Wheeler a famous physicist whose ideas were greatly influenced by his knowledge of quantum mechanics. He has speculated that reality is created by observers in the universe. Despite the fact it sounds very like Bishop Berkley who believed everything existed only in the mind of God, Wheeler was no Johnny come lately. He lived until he was 98 and his credentials make him one of the greatest theorists and a man highly respected amongst the scientific world. One possible explanation he argued was that we are part of a universe that is a work always being created. It is not just the past but also the future that is being made. When we look back, even to the beginning of time, we select one of many possible histories in a multitude of histories. Wheeler argues that it is possible that the universe is a vast pool containing possible histories where the past is not yet fixed. (1)

That does seem stranger than fiction and it is not my argument. I mention it only to show you that there is a wide range of opinion among our greatest scientists, philosophers and thinkers. This range has grown immensely since the advent of Quantum Mechanics entered the debate and there are a lot of controversial and highly imaginative theories doing the rounds. However, they are not to be too readily dismissed as if you read the literature you will recognise we know very little and quantum theory has a very important role to play in furthering our understanding. For example, 'The Holonomic Brain Theory', promoted by Karl Pibram, is a model for human thinking which is very different from the usual ideas. (2) He proposes a model of thinking based on wave interference patterns, situated in the brain. It's complicated and to do with holograms and memory. According to Pibram, waves have a part to play in visual imaging. David Bohm goes further and argues that each of our senses acts as a lens by recognizing or by refocusing wave patterns. (3)

These controversial views are worth mentioning if only to alert the reader to the wider debate involving quantum theory and consciousness. The Physicist Sir Roger Penrose and Stuart Hameroff argue that we cannot ignore the possible effects on conscious of the collapse of the wave function on thought itself. In other words what is going on in the quantum world, which we do not yet understand is so mind blowing that it may have something to do with the working of our brains. If we are to learn more we have to investigate the possibilities and take seriously such ideas and relevant research. An interview with these two can be seen at the following website on Utube. (4)

Remember what is meant by collapse of the wave function - when we look the decision is made, only then is reality created. Penrose believes intuitively that something is wrong with that idea and that a wave function collapse can happen at any time. He accepts that a particle can be in two places at once and believes that the collapse, or reduction, of the wave is an 'objective' event which can be triggered from within and can also be spontaneous. He believes we are not like computers, they cannot think, we can and that is why we must look outside the constraints of mathematics to make progress. He uses the mathematical argument of chap called Godel, that mathematics is not infallible, to support his idea. (5)

A little knowledge about the brain will help the reader understand what an amazing phenomena it is and also acquaint you with some of the terminology. We have a brain made up of about 10 billion neurons or cells and these do a lot of the work. They work on chemical and electrical signals at a few hundred times a second and relay them to the thousands of others they are connected so they are pretty important. All these tiny amounts of energy add up and eventually run down but the brain will still operate very effectively for at least a lifetime and a half. As electric currents are about waves it is easy to see why quantum theory may be seen as having a say in what goes on. Things get more complicated when you add microtubules as there are trillions of them inside the billions of neurons.

Stuart Hofferman has suggested that these trillions of microtubules have an important role to play. These tiny, tiny tubes are found inside neurons, and are seen, by Hofferman and Penrose as the interface of quantum activity. The idea had in the past been heavily criticised but the latest research has beaten off a lot of the criticisms, particularly the argument that the brain was too "warm, wet, and noisy" for quantum to be a factor. The theory has also been helped by the work of Anirban Bandyopadhyay and his team who claim to have identified quantum vibrations inside microtubules. (6)

In the world of biology things are moving very fast and the findings that 'decoherence' is at work in the world of plants and microbes has come as a huge shock to quantum physicists' says Jim Al-khalili in his book. (7) The latest scientific developments in Biology look to be crucial to our understanding. Jim often appears on TV and in his programme Quantum Mechanics, Monday 16Th December 2014 points out as many others now do, that Quantum may underpin life itself. He looks at the evidence that birds which migrate may be able to follow the lines of the earth's magnetism precisely because of the notion of particle entanglement. In the case of a migrating Robin we are told by the biologist it is the only sensible explanation. The right and left hemisphere of the brain are like two entangled particles which ensure the bird is constantly aware when it is going off course. In another experiment he refers to smell and the belief that different molecules fit and form a connection we call smell. This is now a theory under review as it is now thought that quantum entanglement may also be a factor and it is the rhythm or music of the strings which hold particles together which are recognised in the process of smell and memory.

At the end of the programme Jim points to recent notion that entanglement may also be responsible for the miracle of DNA and the double helix. This stems from the work of Elisabeth Rieper at the University of Singapore. The theory is that the double helix does not vibrate and fall apart precisely because entanglement 'ensures the stability of the structure' (8)

The direction being taken by theorists in the search to understand the human brain is inevitable in the light of what we believe we understand about quantum theory. From what is known so far life is a complex building of molecules and all the evidence is that quantum mechanics is a crucial part of that process. It is a law, scientists tell us that is more reliable, more informative and more difficult to understand than any other. That means we cannot ignore the part the strange behaviour of particles and waves and the influence of quantum theory as an essential part of our thinking.

Professor Sebastian Seung a professor at Massachusetts University who specialises in computational neuroscience has written a book called 'Conectomes'. (9) He maps the neuron connections in the brain and believes these are the key to understanding how the brain works. There are billions of connections and you can read about them at his site. In summary he explains about the wires, on and off switches, chemicals, cells, and electrical impulses connecting it all up. They can measure the electrical activity of the brain with EEG (electroencephalograph) and it gives you lots of information of its workings.

I mention these latest developments in order to highlight the fact that if we are to develop an understanding we must start by accepting that the human brain is very crucially a part of the sub atomic world and as such must inevitably be part of what we call the quantum world. We should not therefore rule out the real possibility that spooky action is an unavoidable reality which impacts on our experiences. If that is true then it also follows from this that our present understanding of how the brain works is very simplistic and we have a long way to go before we obtain a realistic understanding of what exactly our experience is – or what the heck is going on!

In looking at the brain we are unavoidably taking on the task of understanding how a complex network made up of billions of active connections is converted into ideas. This surely involves seeing the brain as part of the quantum world. Once we do that I believe we have made a massive leap into the future and have started down a road which will radically alter everything we feel we know about the universe. The following is a quote from The Spooky World of Quantum Biology:

Quantum-level processes within a cell using high-speed lasers, follow the movement of light energy through a photosynthetic bacterial cell, see energy traveling along every possible direction at the same time. It is now very clear that quantum phenomena occur in living systems and must be studied as such'. (10)

On a recent radio 4 show called 'The Infinite Monkey Cage' it was pointed out by the panel just how amazing photosynthesis was. In order to turn sunlight into chemical energy electrons have to be pulled off atoms inside the molecules. It was once thought it then bounced around and somehow brought photosynthesis about. However, they now know that what happens is the electrons do not bounce around rather they use every available route simultaneously and then 'travel along every available route' (11) before they chose the best route without any loss of time or energy in the race to get to the reaction centre, just as they use both slits in the double split experiment.

If that is true about photosynthesis then it makes the current theory about our ability to see by comparing it with a camera look pretty tame. At a basic level we are told the way in which the eye and the brain react to light is just like a camera. Images are projected through the human eye and then projected onto the retina. Then the brain corrects these images so they appear the right way up and we can make sense of them.

Until recently it was all seen as very much a linear system involving electrical and chemical processes. Once the light reaches the retina it is changed, chopped up and sent out again to different parts of the brain. That's when the neurons do their job and we are able to make the different discriminations of size, shape and colour. (12) This is no longer a widely held view in the world of physics and biology. It is pretty clear that whatever the sense, touch, sight, smell, taste or hearing they all appear to be processed by a much more complex system which almost certainly involves quantum theory.

These ideas are now taken very seriously and Stanford University have just published a piece of research, June 2015, which supports the idea of quantum activity in the human brain. It is presaged by a statement from Jonas Richiardi: "Past neuroimaging studies have defined several 'functional networks' in which remote regions of the brain appear to operate in synchrony. A new study from Stanford provides the molecular underpinnings for this theory."

The study provides evidence that parts of the brain are not reliant on a simple chemical or electrical network to talk to each other. Different parts of the brain appear to operate in some other way that still allows them to be 'tightly coupled'. (13)

On the basis of the study Theoretical Physicist George Rajna has submitted a paper to Academia.edu titled The Genetic Basis of Brain Networks. He refers to the study and suggests that it indicates there maybe something quantum like about the way parts of the brain work together and this is related to 'Spooky Action at a Distance'. That means entanglement is crucial to the way it works. In short, things do not operate in a simple linear fashion but, on the basis of this research, there appears to be another way in which parts of the brain are in communication with each other. (14)

There are five basic senses we all know about but many of us believe there is another sense called the sixth sense. This is certainly true of some animals, especially birds who use the gravitational fields of the earth to find their way. Some think it true of humanity too as there are lots of strange things happening in the brain we have no theories for. We all have experience of the moment we look at somebody and they turn around to look directly at us. Most people believe this is because of either periphery vision or some other clue we are unaware of working on our subconscious minds.

Looking for the role of quantum theory may lead to an understanding of many of the unanswered questions about the brain, and what we refer to as the sixth sense. There is a mass of research taking place and most of it has been thrown up in our own lifetimes. I watched a TV series narrated by Morgan Freeman called Through the Wormhole which showed the work of Darryl Bems professor of psychology at the Cornell University conducted a large number of tests on people to see if the subconscious was a factor when guessing. (15) Students sit opposite a screen and are confronted by two curtains. They are asked to guess which curtain has a picture behind it. The computer waits until they have chosen a curtain and only then does it choose where to show a picture. The results are as expected to be around the fifty percent mark. However, this changes when the pictures produced are erotic as the students get far more guesses right than chance allows. It is a small difference of 3% but in the field of statistics and randomness it is apparently very significant.

We are also introduced to the work of Professor Beatrice de Gelder. She demonstrates how studies of blind people who have had their visual cortex damaged are able to imitate the emotions of others seen on a screen – most are blind in one eye. It appears that the subconscious picks up signals if they are loaded with emotions. She goes on to argue that through MRI research she has found that instead of the signals going from the optic nerve through the visual cortex, which in these cases is damaged, they are diverted to other parts of the brain. In short they are not seen but sensed by people who are blind. As such they are the produce of the subconscious. Research in this area is very significant and has clearly established that the results rule out chance guesses.

In the show another researcher, Roger Nelson, talks about 'morphic' fields and suggest there is a Global Mind. He points out that it has been known for a long time that people sitting near random number counters can affect the machine. For the last 30 years he has collected random number data from more than 340 machines around the world. The data is fed into a server at Princetown University and there they are able to record whenever there is a deviation from the norm. He then shows us the graph from the presidential elections of 1998 and demonstrates a massive upsurge in data throughout the period of the president's victory which appeared to start moments before the event. The odds against this happening were a thousand to one - and if you add up the data over the twelve year they have been recording he calculates it is a billion to one. Changes, he argues, undoubtedly occur when there is a big emotional event. One of his most significant results relate to the 9/11 twin towers disaster in America. On the morning of the attack the data indicating a big event came pouring in hours before it actually happened.

Some of the research will leave you sceptical like me but if it is confirmed through more research we can no longer ignore it.

Roger Nelson, also in the same TV show, is not alone in thinking the future can be anticipated. A leading scientist at the institute for 'noetic' science, Dean Rayden also thinks people can anticipate the future and argues it is not always a coincidence which alerts us to possible events. His experiment is simple and involves a person looking at images on a screen some of which are emotionally charged. He finds that if the picture is going to be emotional the respondent reacts five seconds before they see the image. He calls the result a 'presentiment'. There are other experiments in the documentary all showing the wide variety of research in the area of consciousness.

At this point I must mention Arthur| Koestler, the Author of The Roots of Coincidence, as he would be delighted with such results. He was strongly of the opinion that telepathy was an important element in human thought and communication but not yet fully understood.

The challenge to understand the brain has always been there but scientists now think there is far more to it than previously imagined and they have responded with a new energy. Today I read an article by Ian Sample in The Guardian which mentions the work of Jeff Lichtman a researcher at

Harvard University. He believes that chopping up the 8.5 billion neurons in the brain and finding out how they communicate with each other is achievable. (16) Other neuroscientists think that this is reaching beyond the possible and see it as like trying to understand the makeup of every snowflake instead of the weather. Lichtman's answer is; "If you want to understand a city and didn't understand people, it'd be a mystery". I agree and take this opportunity to point out that there is a real possibility that what happens at the sub atomic level is as yet far beyond our understanding but we have to be prepared to acknowledge that the quantum world suggests we can and must allow for something more incredible than simple one dimensional facts based on a fixed and immutable paradigm of the brain. This is the thinking of the future and is fundamental to any understanding of how we are wired up and what part our brain plays in the interpretation of reality.

There is now a new book out by Johnjoe McFadden, called *Quantum Evolution*, in which he writes about several very interesting experiments which suggest that experiments that suggest certain bacteria behave in unpredictable waves. (17) Once again it is a controversial theory he promotes as it also implies that that evolution may not be the survival of the fittest but instead a choice of how particular genes mutate so as to gain an advantage. For those like me who believe, as mentioned before, that no theory has ever stood the test of time this is a very interesting idea.

Until now many have assumed all events are interrelated, hence the butterfly analogy – the belief that a single butterfly flight has an impact on the world in which we live as all things are connected up. This is seen as a deterministic view for it makes it theoretically possible that if we knew how to put all events together in a mathematical way we could determine what the future was to be and exactly what the past was. The world of quantum changes this forever because it is full of uncertainty and unpredictability and is governed by the uncertainty principle. We have learnt that according to quantum theory a particle like an electron cannot sit still as it is always on the move. That means you can never be certain where it is or where it's going to pop up. It is that unpredictable it can actually tunnel out of situations and appear somewhere else. There is also evidence that this tunnelling is very important in the formation of life itself and clearly has implications for evolutionary theory for, as I mentioned earlier, plants and bacteria use it too.

What does this all mean? Surely if the very simplest of life forms rely on quantum processes then they must also apply to more sophisticated living systems, like us. If that is so and I believe it is then we can now continue our search for answers knowing things are far more complicated than we thought and the freedom we have is beyond measurement and maybe our ability to understand. I have no doubt in my mind that what happens in the quantum world also happens in the brain. The easiest way to imagine this, without a full or even partial understanding of how the brain works, is to assume that the something within the brain, already suggested by Roger Penrose above, can cause the collapse of the wave function and what we call reality is then produced. It is a short step from there to accepting that the brain is something we do not yet begin to understand. I am not alone in these ideas. Just go on the internet and check on the latest scientific papers in relation to the issues.

Many now believe that real consideration should be given to the possibility that thinking about something can influence events. Events ought to happen logically and sequentially, as in cause and effect one thing leads to another. But we all know that that is not the way we experience things. Without the concept of coincidence we could not make sense of it all. If we now apply the theory of quantum to life then we can start to explore the possibility that what we think actually affects the shape of our individual realities. It may also mean that if each of us can shape our reality then we are also shaping the dimension our reality is in.

There is a mass of evidence which clearly demonstrates that over eighty percent of our behaviour is determined by the subconscious. (18)

Life is a moving picture, a seamless vision of events, but the reality is that we see, as Wittgenstein argued, in pictures. There is far too much information to take in and we select from the environment and our brains piece it all together. Science has established that this is the process. In one experiment shown recently on the TV they demonstrated that the strategies people used to catch a moving object, in this case a toy helicopter, were not what they supposed but were rather the subconscious deciding the strategy for them. The world we see is the one our subconscious brain tells us is out there.

Clearly nature programmed us the way we are, but it does not follow logically that it was to be able to interact with the world out there by simply interpreting it. It is entirely possible that that our interrelationship is far more complex and that we actually have a role in the way we interpret the world. It would be an act of conceit at this stage to think we had all the answers and there was little left to discover. We must open out minds to other ways of seeing the world. The dog lying next to me does not see it like me, does not look at me and think what is going on in his head. He evolved with a different set of equipment, a different view.

In fact science now tells us in the latest research just collated that we actually see things differently depending on the light and are able to hold the same colour even when the light changes. (19) I am thinking about this and I am driving along in my car and realize I have enormous amount of information now being picked up by my sense. However I also realize that no matter how much information is out there my brain will sort it out so I can handle it. This is part of our condition, we have mechanisms for presenting reality so we can handle it.

What we are finding out now is that we do not know and we may never know what the great mystery is. If we do get a clue it may be through what the scientific world has alerted us to, which is interference. And in line with the arguments presented in this book that is what happens when coincidences occur.

We are talking about our brains and how they work. That is like looking in the mirror and describing what we see. However, we are missing out the observer, the process that allows it to happen. It's the same when you think of your motives or behaviour. You are observing the effect not the cause. To observe the cause you need to switch from one perspective to another and to hold each perspective in focus. It is through the ideas of quantum that such an action is made possible.

CHAPTER EIGHT
THE BIG BANG THEORY

Without quantum theory and its built in uncertainty there would be no theory of the big bang or big inflation as mathematically it would no longer be credible. Quantum makes such theories more plausible by introducing new possibilities, in particular, the concept of ripples at the beginning of time, or the reason why matter and anti-matter shot off in different directions. However, the fact that quantum theory is understood by nobody and is at present more like a door we have not yet opened and seen what is on the other side makes it a theory in process. Already, its implications for the sub particle world and the Standard Model are being looked at daily and the search for an understanding is promising to be a wonderful and exciting unfolding of mystery upon mystery.

In order to theorise about coincidence we need to look at modern theories about how the universe developed, the way in which quantum theory has been used, and its possible implications. So far we have looked at its implications for the subatomic particle world, biology and the human brain but there is undoubtedly a fundamental part it plays in bringing about the whole universe and keeping it all together. That is why we are now going to look at current theories about the universe in order to help us understand the larger picture and to look at the ways in which coincidence may be a factor. I believe that quantum theory has enormous and unimaginable consequences for a future understanding which will make redundant most of our present theories.

But firstly we need a perspective we can all agree to. Let us agree that mankind has not been around for very long and that what we would like to know is how long we think the earth has been around. Few people think it has been around forever and most think it is more than likely it had a beginning. The question is what sort of beginning. A sudden start, a whimper, a long almost eternal process, an evolvement from another state, it has always been and always was, or it is just a product of our imaginations.

The God issue is a serious one and it is not being argued here but if you want to hang on to it then let us assume that God would surely have left a few clues as to how it all happened. I would also add that in order to search out the truth it is just as unlikely that we will find it through a strictly atheist approach as we are a religious approach. Atheism and religion are both equally capable of leading us up blind alleys by using silly laws to justify their position. For example what is the difference between a law that claims that God made the world in seven days and the Second Law of Thermodynamics that disorder is an inevitable process which started when time started. The first can be seen as very much like a fairy story and the second claims there was order which 'existed' either when time started, in a few milliseconds of the beginning, or until shortly after – take your pick. But do bear in mind that nobody can yet explain what time is just as nobody can comprehend a beginning or an eternity.

The current and most popular theory is 'Inflation Theory' This arises out of the recognition by the scientific world that the idea of The Big Bang is very misleading and we still have a lot to learn. The more up to date belief is that there was an 'event' but it was more like an expansion of space than an explosion. Those who subscribe to the theory recognize its limits and its usefulness when trying to understand how the Universe began. They hold on to the idea that, as a rough estimate, all the matter in the universe was held together in a tiny part of space about the size of pin head, an apple or a square meter - just above what they call a singularity. This meant that all the matter and all the gravity in the universe was caught in this small area and everything would be at an unimaginable temperature. This hot dense area or tiny mass of the universe then expanded. It was fast and so hot in the first quadrillionth of a second that the theorists have come up with the idea that it is possible that a quantum event brought about the conditions that made our universe possible, as did gravity pushing things apart rather than pulling them together. (1)

You may ask why a quantum event and the answer is they do not know. Perhaps we never will because the theory tells us that you cannot predict the way in which the particles behave and there will always be 'broken symmetry'. In other words not everything is symmetrical - sometimes it is just the way the cards fall - a coincidence! Without this uncertainty and room for 'asymmetry' the theory would be meaningless.

The idea that it all started with a Big Bang or a big inflation was kicked off by the theory of relativity. However, once they worked out the maths it indicated that our universe went back to what they called a singularity –a single point in time. The singularity is a mathematical problem as well as a philosophical one as it calculates everything down to a point where we arrive at nothing. As that is not acceptable the theorists have chosen the nearest point to nothing as the best place to start the Big Bang off and argue that it is more likely that everything was compressed into 'almost' nothing. From there it is easy to see how they arrived at the Big Bang as if everything is pressed into such a small space it is going to go off with one mighty roar and that means a lot of heat and acceleration as the universe is released into empty space.

The idea that our universe had a beginning is not in dispute. Something happened that kicked it all off. Objections to the current theory are numerous and among them are the idea that you cannot arrive at conclusions like the Big Bang as Einstein's' relativity and quantum mechanics are two opposite theories: one suggests we can eventually know everything and the other introduces uncertainty. 'We have no right to say the universe started off with a big bang' says Robert Brandemberger a cosmologist at the University of Montreal. He argues that the equations they use were not built for 'extrapolating' backwards in time. A very important and interesting point if only that it once again brings home to us all just how fragile the claims of science are in the theory of everything.   (2)

This inflation started 13.8 billion years ago. They have not got a date for the beginning of the universe but they do have an 'era' and it is thought to have started with the 'planck era' which is the beginning of time and the smallest measureable moment of time. It is based on the idea that time may be discrete or in little packets like a quanta. (3)

Planck Time is arrived at by measuring how light behaves over the shortest possible distance which is the nearest you can get to no-time. It is a measurement that can be treated as a packet or a quanta or a dot. Now try and imagine everything in the universe locked up in this moment of time and you can see where the theory of the Big Bang comes from as everything is going to be that densely packed together, and so incredibly hot, that nothing worked, including the particle world.

The big bang is the source of many of our current theories and has led to a number of ideas. Firstly the notion that there is no empty space as you can't condense empty space into a small ball as big as a pinhead. Also it follows that when the bang went off so did space as it had to expand and if space expands then so does time as it is space-time as Einstein said. The time had to expand too. So both space and time are entities which can expand and contract. That also tells you where our current theory of time came from as it makes time a sequential flow related to space.

Nobody has a clue as to what happened before that, although there are many who would suggest they do but if you search the web you will find that there are few scientists who believe in a one-time only Big Bang theory.

The theory has recently become a bit more ambitious and moved on to the idea of inflation and the term 'Big Bang Inflation'. The event occurred in our patch of the universe but could have also happened in many other parts of the cosmos. This expansion of space, not things rushing into space but space itself, happened so fast they can only guess at the figures but whatever the numbers it is seen by 'inflationist' as so mind bogglingly fast that it was faster than the speed of light. Obviously they do not see going faster as a problem at the beginning as they did not initially have the constraints of Einstein's space-time.

I recently watched a Horizon programme called 'Is Everything we know wrong?' In the programme Professor Max Tegmark, tells us of a problem with the theory is that temperatures are too uniform across the universe. With an explosion you would expect a much lumpier universe. We should not have an almost perfect uniformity. This unpredictable finding shattered the theory for a moment and the problem he says, 'forced us to assume some very contrived conditions' He then goes on to say how much they hate unexplained coincidences. This led to a new theory, Inflation. In other words the Big Bang stopped for a few thousandth of a seconds and that allowed the plasma to gather itself up and obtain a uniform heat. It then took off again but was now able to spread itself evenly across the universe. They do not have a theory for what caused the explosion, which I repeat is not just a Big Bang but a 'Big Bang Inflation'.

One of the implications that is very different from the simple idea of a big bang is that our universe may be just one of a multitude of universes as coincidently there was lots of other Big Bangs. It is also possible that these universes have their own laws which may be very different from ours.

Any scientist worth his salt will tell you that we still have a long way to go before we are close to understanding the origins of the universe. In this chapter I intend to look a little closer at the underlying assumptions, especially as its credibility also now depends heavily on what is going on in the sub-atomic world.

The idea that out of nothing there came an extremely violent event at the very beginning of time gained ground from the moment Edwin Hubble, the American Astronomer, established that there were far more galaxies out there than we imagined and that these galaxies are moving away from us all the time and getting faster and faster. There is now a telescope named after Edwin Hubble which is ultra-sensitive and can pick up the very faintest of images. This telescope has photographed a small area in space over a period of time and this revealed that every dot out there was a galaxy and there was an un-measurable amount of them. The light from these galaxies has taken 13 billion to get here and that some of these galaxies are moving away from us at the speed of light. (4)

This movement is the same from any point in the universe which means that the space between us all is getting bigger and bigger. We know that as things move away the light they give off is red, and blue if coming towards us. It would appear to follow logically that there was at some point a beginning, a starting place.

Although Inflation Theory is the 'in-theory' not everybody is happy with it because if there were many Big Bangs there would surely be some evidence of them if only that they would all be smashing into each other, or we would meet some galaxies coming the other way. If the theory is correct the good news is that it also rules out the idea that there will one day be a contraction and everything will start moving back in towards the centre.

The problem for both theories was they needed something to explain why the big event was correct but the Maths was wrong. Fortunately, Quantum then came to the rescue as this enabled them to speculate that quantum had caused a 'ripple' in the universe a quadrillion second after it kicked off. Inflation theory, unlike the Big Bang theory, does have another more serious problem as they cannot explain what made the universe slow down instead of carrying on expanding forever. That is probably why it is still referred to as Big Bang Inflation.

The idea of a 'Ripple' is crucial even though it means something happened they cannot yet satisfactorily explain. At the moment it is seen as the most plausible explanation, a fortunate 'coincidence' that disturbed the creation of matter itself and made the universe possible. We must also bear in mind that according to this theory the creation of the universe would not have been possible without the existence of two types of matter, the other being anti-matter.

Anti-matter was brought into being by Paul Dirac. (5) It is basically a twin copy of matter and when the two meet up they wipe each other out and radiate energy travelling at the speed of light leaving behind a new sub-atomic world. They cannot figure out exactly what happened at the Big Bang but theorise that its possible ordinary matter came out on top. In fact some think there may actually be no anti matter left in the universe but what is amazing is that they can produce it in atomic atom colliders like those at CERN.

Anti-matter is a now a basic fact in the world of physics and is used daily across the world in all modern hospitals for medical imaging in a process called (PET). They actually create anti matter in the laboratory although it does not last more than a few seconds as it is annihilated very quickly once meeting up with matter. Antimatter is used in medical imaging. They inject your body with a molecule which gives out antimatter. This antimatter is killed off immediately it meets up with matter but as it dies it sends a message of where the action is in the form of rays of gamma protons.

We are fortunate that in the beginning matter and anti-matter did not annihilate each other. If they had then there would have been nothing left of the universe. Fortunately, and coincidently, quantum theory ensured they sped off in different directions. So how do you keep the little fellows apart and from killing each other? At CERN they managed to get round this by capturing the antimatter atoms in a vacuum. (6)

The latest news about 'ripples' was recently the main headlines across the world. They theory claimed to be supported by some scientific evidence. (7) They were wrong, but I will explain what happened. They have a telescope at the South Pole looking at the microwave background they believe was left behind after the 'beginning' out there somewhere. The telescope is called BICEP2 (Background Imaging of Cosmic Extragalactic Polarization). The study is based on the idea that gravitational waves were produced at the beginning of time and are tiny disturbances which spread out across the universe. The above article puts it more scientifically they are 'primordial undulations that propagate across the cosmos' (microwave background). The scientific world had grave doubts about this and a few weeks after this article appeared were proved right and the word went out that their finding had been 'Reduced to Dust', (8)

The thoroughness of the scientific method is of monumental importance and in this case was resolved when two different teams with different information, using different telescopes, got together and discovered there was no Noble prize to be had as their figures cancelled each other out.

The theory about wriggly splayed out waves is only possible because of Einstein's theory about gravity and is worth looking at if only because it shows the ways in which scientists sometimes arrive at their theories. But firstly we need to look at the evidence provided by the Planck Team. They had been involved in the task of producing a map of the universe showing the distribution of the microwave background across the universe with WMAP in 2001. (9) These waves are hard to find, exist just above zero temperature and are just like radio or light waves and fill the whole area we call space. The main difference between the different waves is the length of their wave or wiggly bit. The theory is that these wriggly bits exist as curly patterns from polarised light created after the Big Bang, and are caused by a quantum ripple at the beginning of the universe - excellently explained in a Horizon Programme. (10)

The fact that there is gravity is not in dispute as we all feel it acting upon us every moment of the day. That is why the search for the 'graviton' goes on even though no evidence for it has been found yet. Gravitational waves are very different to other waves as, according to Einstein, gravity can distort space and time, and matter itself. That means they can give a waves a distinctive pattern, or 'wriggly bit' (11). To understand what they mean by this we can make an analogy with gravities effects on tidal forces. When the moon is overhead it pulls the tide up towards the moon and also distorts the sideways movement as if it is pushing the water sideways and that's when you get your high and low tides. Matter, in this case water, is stretched and distorted. Since Einstein educated us about space time, the notion that space and time are effected by the mass of the huge objects in space, we now see space as being akin to a rubber sheet or 'almost' with similar potential to a lump of dough – you can move bits about but you retain the same mass. In other words gravity can change things so much so that the nearest point from A to B will not be a straight line but a bent one. That also implies that if we shake space up a bit we can leave footprints or evidence of events we can use as evidence of our theories. These can be marks or wriggles. There will be many of them and that means patterns we can pick up and interpret. It can be likened to being hit by a tidal force and stretched in one direction and squeezed in the other. In other words you have been hit by a gravitational wave. Boom, boom!

From this theory about gravity and how it behaves science has been looking for evidence of this stretching and squeezing on early light from the universe out in deepest space. That means light that has been travelling towards us since the big bang and is the stretched out remnants from the beginning of 'our' universe. We can't see these waves but the theory is that when the universe expanded these first waves were stretched and expanded into micro waves and radio waves called cosmic microwave background. This fills every part of the universe and has been travelling towards us since the universe began. This background is from the beginning of time when everything was super-hot, so hot and so momentous an event that strange and unpredictable things happened.

We can see the logic of this train of thought which leads very neatly to the following speculation: 'In a fraction of a second after the Big Bang, the explosion of space-time that began the universe 13.8 billion years ago, the infant cosmos inflated to many times its initial size in less than a quadrillionth of a second.' (12) Most scientists are waiting for confirmation and it will be some time yet and loads of experiments before it is fully, if ever, accepted by the world of science.

Many believe that if there was a big bang, now called Inflation, and other galaxies are all moving away from us then it would appear safe to assume that any other big bangs occurred far out and way beyond our universe. Please note that this does not rule out the possibility that our Big Bang occupies its own space and time. In other words there could be others in their own space time. After all, the really big question we have not answered is what is time and how is time possible? This too is something we will look at in another chapter.

To assume it all started with a big bang is too simplistic. Few now believe that before the Big Bang there was nothing for the simple reason it does not make sense. It's all effect and no cause. For the universe and its life forms to occur there must be an unknown explanation. Something must enable the possibility and amazingly, wonderfully timed and very excitingly, the only known theory which may or may not enable us to at least comprehend where we should start looking is that of quantum theory.

Not everybody agrees with the current perspective of the Big Inflation theory. Gabriele Veneziano works at CERN. He points out that we all wrongly assume they all adhere to the Big Bang theory. That is not the case and he suggests no Big Bang but more of a whimper. In other words it was a slow evolvement up to the moment of the big event. The universe had been around for a long, long time, so long we cannot imagine it even in a quantum computer. He takes up the issue of the singularity as an obstacle to progress and an indication we need an alternative theory. The singularity contradicts Einstein's theory of gravity and relativity as it makes them irrelevant at the beginning of time. That means he was wrong! It would also mean that in the heat and pressure of the beginning gravity would be as strong, if not stronger than all the other forces. That's a problem, but the other problem is, as pointed out earlier, they do not know what gravity is.

Venezianos theory is based on string theory. This suggests that as strings are so unimaginably small, so small they are best seen as taking up no space like musical notes, then the singularity does not arise as a problem. This also means that something existed before the Big Bang. (13)

And so the model survives, big bang, inflation, galaxies emerge expansion slows it down, the universe is stretched. It is all there in the Standard Model of cosmology. Whatever the differences of opinion there is something about the theory which suggests we are looking in the right direction and many of the claims and counter claims will eventually help create a perspective for the future. The search for a beginning is the lot of mankind, an explosion, an eruption or an inversion of time we can only speculate. Whatever, we have to believe there is a history of the event. The best we have so far is the microwave background which may have come from some momentous event like the Big Bang or Big Inflation but may also have come from some other event possibly involving 'time' itself. We will discuss this idea later.

CHAPTER NINE

From Black Holes to Many Worlds Theory.

Quantum Theory has changed the way we think and that is very clear when we look further at what is going on in Cosmology with concepts ranging from Black Holes to Dark Flow. This will mean a little repetition of some of the points I will be making which, apart from helping build an understanding, is itself due to the lack of a theory covering everything and the interconnections of current thinking.

Scientists argue that Dark Matter makes up almost a quarter of the universe. Although we can't see it about us we can see evidence of its existence light years away. It all started when as the universe started to uniformly spread itself out and a quantum ripple, or an event we can't fully work out yet, created the asymmetry making life possible. In other words without asymmetry we would not be here. – The quantum ripple allowed the universe to start clumping together and that's why it all became possible. Easy!

Each of these developments is amazing and need looking at not only to educate ourselves about the ongoing debate in the scientific world but also because they are among the first indications that the world of science is beginning to think there may be more to it.

Black holes are seen as the remains of giant stars which have run out of fuel and collapse in on themselves under the pressure from their own weight. These are known as super nova which collapse into their centre. At that point it becomes a place where scientists and philosophers have no trouble conceding that the laws of physics break down, a strange world of strong gravity where straight lines cease to exist and nothing can resist its pull. In some cases we end up with a mass maybe as little as 20 mile across but so powerful that even light cannot escape its pull. Some, as indicated by todays paper, are '12 million times larger than the sun and is 12.8 billion light years away and formed 900 million years' AFTER THE BIG BANG! The Independent 26th Feb 2015.

Outside the world of media publicity there are now possibly as many theories about these events and observations as there are scientists who accept that as yet we understand very little and need to use our imaginations if we are to move beyond the hackneyed views of the media. The world of cosmology is a huge canvas and with modern telescopes there have been some wonderful discoveries. The irony is that there a number of controversial ones like Dark Matter and Dark Energy which have only served to add to the mystery.

This is not true of the Black Hole as that was theorised by a number of people before evidence of its existence came to light. It started when they noticed a radio emission from an area of space. They watched the area it came from and noticed that over time it covered a passing star. The spot became the first Black Hole and is called Cygnus X-1. The Black Hole lets nothing out including light and is thought to have what is called an event horizon. This is basically the point of no return. This was considered by most people as being a place nothing was able to escape from and as such the end of its existence if it was drawn in. Stephen Hawking argued the point very forcibly even though it contradicted the views of quantum theorists. This was a red rag to a bull for Professor Leonard Susskind who appeared on the Morgan Freeman Series, (1) He takes issue with this argument and believes it is wrong to argue Black Holes destroy all information. That would end everything, not just matter but time and history. It is not comprehensible, nor is it to physics in general as he believes it violates a law of conservation of information. Now this law is basically saying that everything that exists will exist forever as information. It cannot be erased, obliterated, wiped out. It is certainly hard to imagine that is possible, it just defies our common sense to try and think that 'what is never was'.

Fortunately the professor has other ideas which, although radical, get round the problem and save the law of conservation of information. He believes that our three dimensions can be contained in the two dimensions of an event horizon forever. This means we are holograms stored on a two dimensional plane at the edge of the universe. (2)

Leonard Suskind uses as evidence of his theory the light experiment with the two slits. This he argues confirms his view that we live in a two dimensional hologram and that maybe the photon goes through two holes at once until its spotted and then the one hole is closed as something else is using it. Hopefully these ideas have inspired you to think of some of your own.

Of course the universe could be an infinite breathing, frothing mass which goes on and on without singularities and boundaries. In other words there is no Big Bang but rather an ongoing of little bangs and black holes. In a recent TV programme The Observable Universe, Horizon Sept 2012 it was pointed out that most galaxies have a Black Hole. (3) They looked at 90 and found 32 had one. They are extremely confident of this and are just as confident that matter in the form of particles goes in and out of existence on a regular basis.

The spread of knowledge and our exposure to the very latest theories is a wonderful thing. However, just when you think you have caught up with the latest theory along comes another one which throws it all up in the air again. It's a fact that helps me feel more confident about the need to be very cautious about scientific claims. On the 25th January 2014 a reference to a new Stephen Hawking theory pops up in my email. It refers to his latest argument that black holes do not exist in the form we are currently told, nor do they have event horizons – the edge at which nothing can ever return. The event horizon is replaced with an 'apparent horizon'. (4) These hold light and information for a short time and then release it again in some strange and altered form. This happens because there is a form of quantum chaos about a Black-hole in which nothing is predictable, so much so that what comes out is not necessarily in the form it went in. So **Leonard Suskind need not have worried after all.**

This means that instead of an inevitable fall into a black hole which would rip everything apart and from which nothing can ever return we now have a 'highly energetic region'. This new theory is a consequence of the need for a new understanding brought about by the implications of quantum theory. This has brought about some debate in the world of theory as it is forcing people to take sides on the issue. In quantum theory information cannot be destroyed - if you know what went in you can work out what comes out. This is not true if you use Einstein's General Relativity as that states that nothing can escape from an event horizon once it crosses over the edge.

The mystery of space, black holes and gravity is not yet resolved. In fact the mystery deepens almost daily. The latest theories involve the question of Dark Matter, Dark Energy and another chap called Dark Flow. Meanwhile at CERN the search for evidence of these theories continues and with the upgrade this means it is now handling. 6,000,000,000,000,000 bytes a second according to New Scientist. One theory is that as everything started at the big bang then if they produce a big bang at CERN they may see dark matter before it became invisible.

Dark matter was first noted by Vera Rubin in the 1960s. (5) She saw that the stars in the Andromeda galaxy were all rotating far faster than they should be. She worked it out at about 25km per second. At these speeds they ought to have been breaking apart. The fact that they were not could only mean that something was holding them together. Hence the birth of Dark Matter to explain what was happening. It was also noted that space is expanding ever faster, and that the light from parts of the universe may never reach us. An explanation was needed and along came Dark energy. In summary, in order to explain why the planets hold together without breaking apart we have Dark Matter. In order to explain why they are moving apart at great speeds we need Dark Energy. There was also another reason why the theory needed dark-energy to remain a plausible explanation. If at the beginning there was this small dense little area that would eventually explode into our universe it would have basically been no different to a Black Hole. So how did it expand? Inflation and Dark Energy came to the rescue and the theory has survived.

The idea of Dark Matter has led to some great leaps of imagination but the first thing to remember is that it is just a theory and its needed to explain why there is not enough mass in the universe and to explain why the planets and the stars behave as they do and don't fly apart. This is made very clear in a recent TV programme, DANCING IN THE DARK. (6). It introduces a group of eminent professors all involved with different ways of looking for Dark Matter. There are a number of approaches to this. One approach is called the Hooperon after it creator Dan Hooper. It suggests that Dark Matter created the particle world when elements of it crashed into and annihilated each other.

Dark matter is believed to be out there but we are unable to see it. That means the only way we can find it is by looking at its possible effects on planets in its vicinity. Note that there is no question that our current theories are wrong. It is taken for granted they are correct and we just need to find the missing piece. One of these particles possibly responsible for Dark Matter is called is called an 'axiom' and is theorised to have no spin, no charge and very little mass, sounds very close to being as invisible as the matter it represents. Another idea is of the MACHO a massive astronomical object made of a little bit of matter, no radiation and it drifts through space like a jelly fish. These are but two of the ideas being proposed as candidates for Dark Matter.

They even have a figure for how much of the stuff, which cannot be seen, there is in the universe. I repeat, it works out at that our universe is made up of about 23% of dark matter, 72% of dark energy, but amazingly only 5% of the matter we walk about on makes up the whole picture. (7) Dark Matter came along after the theory of the Big Inflation and it has to fit in somewhere if it is to make sense. It is now thought to account for the way gravity holds galaxies together. It is also seen as responsible for messing up fluctuations in the microwave background of the universe at the time of the big bang.

The issue of Dark Matter has not yet been satisfactorily resolved and they are hoping to find it at the upgraded Atom Collider at CERN. However, identifying it is only half the problem as that would tell us nothing about what role it plays in interaction, and surely it does interact, with the rest of the universe. How this invisible phenomenon fits in with nature sometimes being random and ripples in the Big Bang occurring is still a mystery. In fact Matt Strassler, among others, who works at CERN states on his web site that they are not even sure dark matter exists. (8)

Next in line is Dark-energy. This is a highly respected theory and is used to explain why the distant stars were becoming more and more distant and moving further away than was expected. The discoverers of this information were given the Nobel prize in physics in 2011.Nobody can tell us what dark energy is but one suspect is named as the cosmological constant. This is seen as an unchanging energy which might emerge from the froth of short-lived, virtual particles that according to quantum theory are fizzing about constantly in otherwise empty space. Virtual particles are temporary particles that continually form, disappear and reform. (9)

The other theory is that we have the wrong theory of gravity and some researchers now argue that in Dark Energy we are seeing an entirely new side to gravity, which out there at the edge of the universe light years away it possibly changes from an attractive to a repulsive force. If true this would mean dropping Einstein's general theory of relativity in which he see gravity as responsible for the bending of space and time, and up to now has accurately predicted the motions of planets throughout the solar system.

It is now very clear to the world of science that normal matter, the thing everything is made of, is only a fraction of the universe and is outweighed by Dark Matter. This in turn is very much outweighed and surrounded by Dark-energy. It was once thought that as the galaxy expands then it will be slowed down by the pull of the dark matter and eventual we will all end up crushed in a small ball. However, latest research shows that this is not going to happen and the movement of the stars in the outer rim of the galaxy is going to continue to move away. So the universe appears to be defying gravity and the reason is that there is another invisible mass called Dark-energy and that is doing the pushing. Dark Matter pulls and Dark-energy pushes. They are also playing with the idea that as the universe expands then so does Dark-energy.

There is a long way to go before we are able to get a fuller understanding of what is going on out there in the rest of the universe. However, the strides they have made and the brilliance and originality of the research taking place is phenomenal. In order to study the behaviour of anti-matter to discover why nature prefers matter to ant-matter scientist have experimented with anti-hydrogen atoms. A hydrogen atom is an electron orbiting a nucleus. They excite the atoms by firing light at them. Once they relax the atoms emit light and this provides the experiment with a spectrum, or pattern. They are assuming that the anti-hydrogen atom will have the same spectrum as hydrogen –if it doesn't they will still learn something. These are wonderfully imaginative experiments which may one day show us the direction we should be looking in.

I argued at the beginning of this chapter that for a beginning, the moment in nothingness, to occur, we have to imagine that something came out of nothing. It is impossible to imagine nothingness but let's suppose there is such a thing and out of it appears something. Now it may well be that is in fact the process but if it is it also follows that it is something which happens again and again and in the vastness of eternity has undoubtedly happened many times before and since. However, if there was always something then the question changes and we should be asking what is it that was always there. It could be dark matter which we know nothing about, or dark energy, light or waves of which as evidenced by the quantum world, we also know little about. What we all appear to agree on is that it was not matter which was always there as that is a product of the thing we do not know about.

Before venturing further into the unknown we need to remember that no theory has ever stood the test of time without being changed, plus the recognition that the human race is only at the beginning of the journey towards an understanding. This latter point was made clear on a Todays Radio4 program chaired by Melvin Bragg. He talked to an august body of scientists Anne Green, Carlos Frank and Carolin Crawford. (10) The panel reviewed 'Dark Matter' and it importance to the big bang. They confirm much of what I have mentioned so far in this chapter. Science was fairly happy with the big bang theory until astronomers put a spanner in the works that galaxies were moving too fast. They ought to be moving at different speeds and were all behaving badly and were messing up the theory. It is now argued that the Big Bang predictions do not work and that 'Inflation' is a better idea and that if there is inflation, it goes on forever and everywhere. It appears to me that it is only the belief that there was nothing and then something that holds us back and makes idiots of us all.

The Radio4 team then followed this up with a look a Dark Matter and went on to explain that there is no visible evidence of Dark Matter. It was invented to explain why the planets are not moving as gravity should predict - although they don't know what gravity is yet. They don't even know if Dark Matter is made of particles. It is a theory based on the observation of speeding galaxies and 'Lensing'. This is another ingenious experiment which I mentioned earlier. Lensing occurs because space is bent by gravity and they are able to observe light traversing the universe and following the curvature of space caused by the planets. However there is not enough gravity according to the mathematics so something else is causing the lensing which they see as dark matter. The best theory they have of what it is are called weakly interacting massive particles – WIMPs for short. (11) You may note when listening to Media-speak that they do it with absolute conviction even when the concede that they are excited about the possibility of new findings – suggesting they will find a way of fitting it in to existing theory even if they have to hit it with a hammer.

Add it all together and like me you have to leave room for humility and awe at how little we know and how much we may have yet to learn, especially if we keep in mind that the universe is all of it from as far as we can imagine to as small as we can imagine too. The Big Bang Inflation is a comforting idea which tries to explain how it all began, but most of us recognise that it may not be possible for us to explain it at all as it is so awesome a task. Maybe there was a beginning, but it is also just as feasible that there was another beginning and then this one. It is impossible to think there was nothing and then something came out of nothing. It seems to me that it is beyond our ability. The only way I think we can move on is to assume that there was always something and then there was us.

It is far more pleasing and interesting to think there was something else and something else not comprehensible in this, our current universe. There is possibly another dimension we know nothing about but are now able to obtain clues through the amazing experience of the world of quantum mechanics

Clearly while we are looking for a theory of everything we must also recognize that there may be no possibility of ever arriving at one, as we are part of the very thing we are trying to understand. The existence of our little fingers has the same roots as the largest of black holes.

The notion that the universe is an organized symmetrical construct which can be taken apart and simply reconstructed through mathematical formulas is history. Science has moved on since Einstein said 'God does not play dice' and it is now a key part of current theory in quantum mechanics that nature is both random and organized.

At every level, either in the sub atomic world or here on planet earth, it can be clearly demonstrated that the slightest of changes in the balances of forces and none of us, not even the universe, would exist. It the same with gravity if it was any weaker or any stronger than it is that too would finish us all of. Everything appears to be so coincidentally just right. Fortuitously, they knew this long before they knew about the other 95% of the universe. This is not to say things will not change again –they will, although we may not be around when they do.

We must accept that our universe has been around for an awful long time and produced billions of stars and planets and the possibility of life. That last phrase fits in perfectly as a common sense assumption. But even that is a puzzle. We must assume logically that life giving situations have occurred before now and there are life forms all about the universe. We also note that we have a tendency to invent what we imagine and we would not be too surprised if one day inter stellar travel is a real possibility, as it surely would be if other life forms had emerged earlier than us. So where is the evidence, why has nobody been in touch? Are we or are we not the only advanced life forms in the universe, or are there others? It's known as the Fermi question; 'where is everybody?'

There are a few recent new comers to the list of maybes we need to mention as they are the result of new observations which may be important. One theory claims that the universe owes its beginning to, Galactic Polar Jets. (12) This theory is all about huge jets of matter emanating from Black Holes. Another potential source for the universe, but it does not yet have the status of the one big event. I picked up the second in a recent TV Horizon program. It was proposed by the astronomer and cosmologist Sasha Kashlinsky. He suggested that some galaxies out in deep space indicated another addition to our lack of understanding called 'Dark Flow'. He believes there are other galaxies with different laws that are moving away from us in ways we do not understand (13). Cosmologist Laura Mersini-Houghton who I mentioned earlier, believes that Sasha Kashlinsky's findings are the "most straightforward indication of the existence of the multiverse."

Big Bang, Inflation, Black-holes, Galactic-Polar-Jets, Dark-Matter, Dark-Energy, Dark-Flow and none of it fits together except to tell us we have a long way to go in the search for a full and satisfying understanding.

CHAPTER TEN

WHAT A WONDER WE ARE

You will never change things by fighting the existing reality. To change something, build a new model that makes the existing model obsolete. Buckminster Fuller (1)

So there we are, if we understood it all we could set about trying to create it. That would make us God-like. I started off using the notion of coincidence as a means of opening up the debate, partly for myself but also for all the non-scientists amongst us, on the nature of reality. At the time it was simply a device which I thought would help in understanding the lack of knowledge most of us have about the goings on in the world of science. As somebody who had stayed abreast of the latest theories over the last fifty years, and long before the media had started to make such information available to all, I felt I had an understanding not generally available to the non-scientist. It is now good to see that thanks to the speedy development of the media and the digital world such information is now available to all. And so it is with great joy and a sense of vindication that I note that many others, especially in the world of science are now coming to similar conclusions, which, in summary, mean we really know very little at all.

The good news is that due to the development of the internet and other media there is now an honesty about the fact we know little. In fact I was inspired today myself by an article which epitomizes this development. It appeared in the Science Section of the local paper. I think it also helps to put the production of theories into perspective.

'A strange signal caught by an Australian telescope – which had baffled astrophysicists for 17 years - turned out to be generated by the office microwave.' (2)

Having read as much as I can and given it some serious thought I am now firmly of the belief that coincidence is telling us something about the nature of reality and only when we find out what that is will we be able to further understand what is really going on. My belief is based on the knowledge we have about what happens in the quantum world. It is now clear that we have to treat events in this subatomic world as being brought about by the actions of both waves and particles. This is now a statement of fact and one that cannot yet be satisfactorily explained by current theories. The world of science and scientific theory has been opened up and found wanting. There are, it is said, 'now as many theories as theorist.'

This does not mean we throw the baby out with the bath water. The sensible thing is to treat the Standard Model as a paradigm, a tool to continue our search for a fuller understanding of the sub atomic world. However, we must now accept that we are on the threshold of new and exciting times and that current models will almost certainly be drastically modified in the near future. This new perspective also allows us to treat a number of alternative theories as serious contributions to the way forward.

This is not to accept everything uncritically as I too find it hard to believe that we can ever resign ourselves to the idea that it all depends on whether or not we are 'looking' at or measuring the particle world. Looking is certainly consciousness creating an event but not the creation of matter. There is more to it than that and fortunately the notion that it is consciousness itself which determines the behavior of a particle is very much under the microscope, especially in the world of biology. It is now suggested that noise itself is a factor in initiating the event in which a wave becomes a particle. Noise, as a result of any activity, appears to be a key factor in the book 'Life on the Edge'. They argue that you do not need to look or use a measuring device to turn a wave into a particle as the 'the atoms and molecules – undergo thermal vibrations and get buffeted around by all the surrounding atoms and molecules so that their wave like coherence is lost'. (3) That, they argue, is enough to ensure the event. That is also why any experiments have to be cooled down to almost absolute zero to avoid the problem. They quote a number of experiments including that of Graham Fleming and the idea of 'quantum beats'. (4)

Noise is anything that causes anything in the sub world, it's the hurly burly of wave and particle activity, something that passes between any two points and as such provides a measurement. We have to remember that this activity takes place in a world that is hard to imagine but is like a grain of sand in a desert the size of the earth. It is a very active busy area in which jostling and crowding and coming and going are happening all the time.

This area of activity in the sub atomic world is as mysterious as consciousness itself. Let us probe the issue a little further by going over some previous ground asking, what on earth is a wave? To be a wave you need to move in something and as we now know, thanks to the world of science, there is no such thing as nothing. This leads to the question, what then are waves moving in? The answer appears to be 'fields!' A field is likened to a big wave or lots of waves all joined together and constantly on the move and constantly in a state of flux. However, we still have to ask what they are all moving about in or existing in. So far they appear to be a mathematical construct and difficult to explain – as you saw with the Higgs Boson.

We should not be too surprised at this idea of fields and constant change as, just like in our own experience of reality, it is a common event in the particle world – remember nothing lasts forever, everything changes into something else and even if it didn't 'chaos theory' the science of the unpredictable, would ensure the change. (5)

If there is no such thing as an empty space then the only way to get movement, and therefore time, is by an exchange of matter – you take my seat I will take yours. I was unable to find a scientist who still believes in an empty space. In fact there is a recent theory that 'space' exists in 'chunks'. This is a theory reported in the Guardian Newspaper (6). It is an idea put together by Craig Hogan a theoretical astrophysicist at Fermi lab. He questions the idea that space is infinitely divisible and asks where does gravity fit into the scheme of things? Craig proposes a 'quantum' of space or as the article states a 'chunky-bit'. He is currently conducting an experiment which he hopes will make things clearer. This experiment is great news as it may have the unintended consequence of confirming something new.

The following experiments suggest it is not only that change is constant but also there is something else at work which has to impact on any understand, or any theory we may suppose about reality. Scientific experiments have clearly demonstrated this by creating an opportunity for the photon to behave as a wave or a particle. Amazingly it has no trouble deciding on being one or the other and this has been very clearly established, with numerous experiments, that the photon is able to go one way as a particle and the other as a wave. How do the photons decide when to tunnel or not to tunnel, what is the determining factor? Also, and just as surprising, it equalises itself, implying that it knows which side to go through next in order to ensure that both sides have equal amounts of photons.

Let me remind you about tunnelling. Jim Al Khalili in Life on the Edge, page 10, tells us that tunnelling is necessary for the production of light. Fusion has to take place in an atom for light to occur which means two particles have to get together in order to release energy. However, they find this getting together difficult as the closer they get the greater the force pushing them apart. They only way to do it is to become a wave, drop into position next to your opposite number and wham bang wallop you get a blast of light.

The latest development is at DESY a particle accelerator in Germany. They are taking it one step further by looking to see what allows photons to change into particles, then into waves and then back into particles again. The whole process is called tunnelling and the particle they are hoping to discover is called is called a 'WISP'. (7)

I must also repeat, as I stated earlier, some particles decay in less than a second, but some appear to live a lot longer, and nobody knows why it happens. The ones that do not decay, like the electron, are seen as 'stable'. However, if the 'stable' electron meets a positron (Anti electron) they annihilate each other and are turned into pure energy. It appears to be the security of a good home inside the atomic nucleus which keeps the particles relatively stable. Outside they appear to be at the mercy of the elements. The interesting thing is that whatever the process any and everything can be changed into something else either by an instant and spontaneous process which cannot be predicted or by the awesome capture of a black hole, a big bang, a collision or, if you are really imaginative, a quantum ripple in the universe.

In other situations it is pointed out that particles like photons get bullied by electrons all the time and often have their wavelengths changed so that they end up as two photons instead of one. That is because of the law of conservation again. In other words the energy of the two has to be the same as it was originally. However, that means they are now entangled, joined at the hip and it is spontaneous! It is now pretty clear that nothing is ever completely annihilated and that there is clearly an interaction between things. Things change and move and change and there is a constant supply despite the fact that everything changes.

There are numerous theories about this, and the idea that the underlying activity, or noise, of the sub atomic world plays such a significant role, appeals to me. However, we must not forget that all of these theories start with what we now know, and accept, that the scientific evidence that a particle can be in two places at once, even though they are a million miles apart, and that communication between entangled particles is spontaneous. There is no empty space, nothing is lost, and things just change into something else. This is true even at the sub atomic level. Matter or energy cannot be lost, only changed. Too many protons or too many neutrons and you do not get chucked out of the nucleus you get changed into one or the other until the balance is restored.

Remember, there is no empty space just change. Things change and become something else that in turn changes as it too is in a process of change. It's the scientific law of conservation. You may ask why? The answer is that it is the law! I offer a silly example of this law but it puts it into an interesting perspective. Imagine that in the world of the microbe, whatever the gender potential, when there are more in the group of one sex or the other, change occurs and either one or other of the sexes will change sex in order to restore the balance. That is exactly what it is deemed to happen in the particle world. This law of conservation and its importance cannot be underestimated. It's the law. Let's hope we are on the right track.

.

CHAPTER ELEVEN

THE OTHER DIMENSION

Hopefully by now the reader is aware that nothing has been settled and there is lots of room for alternative theories about the nature of reality. Quantum entanglement and time are the two issues we know little about and I believe that they are the key to any further understanding, especially the relationship with coincidence.

Entanglement, Einstein's spooky action at a distance, is the greatest discovery of our age which I believe opens a door into a future understanding of reality. I am also of the belief that what we call coincidence may in some way be part of its strangeness too. It has massive implications. It implies there are possibly at least two dimensions! It is two events, incidents, happenings, bits of the sub-atomic world linked together. One is part of our universe and the other out there and separate but connected up to us in some way. It is logical to assume that entanglement occurs because something is at work which links things together across time. It is not very hard to imagine that one photon relates to one dimension and the other to another dimension. In fact this idea is fundamental to the multi-world theory of Everett and Wallace and String Theory, less ambitiously, has suggested there are ten dimensions.

Many believe there are a multitude of other universes running parallel to ours. David Deutsch has no doubt and sees us as living in a 'multiverse'. He very convincingly uses the two slit experiment to explain why. His conclusion is summed up in the statement:

"The heart of the argument is that a single-particle interference phenomena unequivocally rules out the possibility that the tangible universe around us is all that exists."

If you read about the experiments in the world of science you will have little choice but to take his arguments seriously. (1)

Events in the sub-atomic world and mathematics indicate such theories may possibly be correct. And that is why we do not want to lose sight of one of the most important of the events that have inspired these ideas, entanglement or 'spooky action'. This fact of life is crucial to a future understanding. Let me remind you. If they measure the state of the first photon and immediately measure the state of the second photon they find that the first measurement affected the other, despite the distance between the two. It is seen as being spontaneous and that is only going to be the case if they are joined at the hip and living apart at the same time and faster than light activity is taking place.

The idea that two parallel events can take place outside of space and time in an instantaneous flash is not explainable within our current theories and you may choose to think it as nonsense. So far it is only the idea that there may be other dimensions which provides a possible solution. It is the logical conclusion of a lot of scientific observation which we must explore.

The scientific interpretation is that this is a case of a quantum collapse where the measurement of one state forces the collapse of the other state. It means that in the end two photons are in some way connected, and it does not matter how far apart from each other they are. It also means that there is unavoidable faster than light action taking place. This is very serious science and has been proved to occur in experiment after experiment. In fact Nicholas Gisin showed spooky action at speeds of 10,000 times the speed of light and only stopped there because he did not have the means of measuring it any faster. And this is a scientist who:

"Despite his critical work….Gisin has been active in searching for alternative mathematical formulations of quantum theory, especially ones that might replace the ad hoc assumption of wave

functions "collapsing" when measurements are made". (2)

You may think that these things cannot be true as they defy logic and are very difficult to understand - although I believe one day we will. Everything I have said can be confirmed on your computer, just check it out by typing in the name, the experiment or the TV programme referred to. Hopefully you have reached the point where you are beginning to think something else must be at work.

Let me remind you of the mystery again. Waves appear to have a number of attributes, they can traverse the universe and become particles and they can also turn back into waves and then become particles again. The puzzle is that at any time any part of a radiating wave can be transformed into a particle. It is not inconceivable that the wave operates under timeless laws but it becomes a piece of history, or time, as a particle. Add to this the knowledge that research has clearly demonstrated that there are two things that are so far incomprehensible. One is entanglement and the other is tunnelling, the latter we now know is also involved in photosynthesis. They are serious comments on what is happening in the sub atomic world within and below the level of the particle zoo structure being presented to us through The Standard Model. We can go no further until we develop a theory to explain them.

Every great scientist alive today will litter their writings with the phrase, 'We do not know'. This is especially true of entanglement. It has been established as a fact for a long time and will be central to any future theory about reality. You can search the world of science but you will not find a simple answer. There is a spontaneous connection between two particles and we need look at what we means because I believe it is telling us something about 'time', and most obviously, that there is no history to 'spontaneous' in our current, nebulous, concept of time!

Further on you will also discover that based on evidence like entanglement, the two slit experiment and tunnelling, waves do not appear to have the same constraints in time as the rest of the particle world. It is not inconceivable that the wave operate under timeless laws but it becomes a piece of history as a particle.

It is impossible for us to imagine empty space without something in it. Once we think of something in it we have created the conditions for space time and movement even if that movement was just the imagined movement between point A and point B. Time is movement whether in a Black Hole, an adrenalin filled moment, or an event in the sub-atomic world out there. Nothing exists until there is movement. It is the answer to 'Zeno's Arrow Paradox' (3) The answer is a recognition of the idea that you cannot have 'still' and 'not-still', only one and not the other. When we think of no movement we are standing outside time and using dumb numbers which on their own tell us nothing. We are trying to reduce time to nothing, but time always occurs between two points and is in fact only comprehensible as movement.

Time, like infinity, is not really understood and just as we cannot imagine infinity so it follows we will get nowhere calling time infinity. Einstein says the following about time: "it is an illusion if forever persistent" He saw it as relative and able to go faster or slower depending on our movement through space. He also added that as an illusion it would make past present and future one and the same thing.

The implication of that is that what we call 'now' is a construct of the brain which allows us to make sense out of our individual experience. Einstein put time and space together and called it 'space-time'. He recognized that you cannot have space without time or time without space. It also means that we cannot have movement without space-time as nothing would exist. However, I think that the real illusion is that we think of space time and movement as three separate things. Space, time and movement are all one and the same thing. They cannot, in current theory be separated except with the consciousness, and therein lies the illusion.

It is often thought that movement in space-time may not necessarily be between two meaningless points like A and B. Movement may actually be between two real units of time sometimes referred to as quanta. If that was the case it would also follow logically that these units would somehow be able to relate to each other, and pass time on and across each other. But that would involve a 'moment' outside these units of time. We can cope with the idea that spontaneity can happen between two units of time but not with the idea that movement can take place between two units of time as that requires a go between or another unit, like the undiscovered graviton. It is hard to imagine an event independent of time which can move from one unit of itself to another without even a threshold to cross. That is not possible if everything is subjected to the rule of time. All this makes movement or the time we know and recognise a different animal to everything else in the universe. Later on I want to propose a different theory of time but I make the point above to because I want to suggest the idea that there are indeed two different aspects of time we do not yet understand.

I like many others do not believe that time is an illusion. The theory I propose is that there is real non-illusory time and a point, as the evidence suggests, where time behaves differently. This different behaviour where time is thought to go backwards is no longer just a theoretical claim of

Richard Feynman, John Gribbin and others but has now been established. It was achieved in a recent experiment by an Australian team and published in May 25 2015 in Nature Physics. It confirms the hypothesis through a complex experiment and states quite clearly that future events can indeed affect the past. (4) It appears to me that there are two different times which co-exist and the time we know and are familiar with is just part of one big cake.

There is some way to go before we truly understand time, which leaves us with a choice: time is an illusion, time exists, or time exists and there are other times. I believe all three and find that in order to think about the subject it is useful to see time as being first of all 'real'.

The idea that 'time' actual exists out there in reality is inspired by the work of Sandu Popescu and his team at Bristol University. (5) They have a theory that time is the result of the particle world seeking to merge with its environment. Entanglement in the particle world, they suggest, accounts for the way energy is spread about and interacts as it seeks an 'equilibrium' with its environment. Tony Short, another quantum physicist at Bristol uses the example of a cup of coffee to explain: 'entanglement builds up between the state of the coffee cup and the state of the room'. *Objects reach equilibrium, or a state of uniform energy distribution, within an infinite amount of time by becoming quantum mechanically entangled with their surroundings.*

This happens because the relationship between the sub- atomic world and the universe is unreliable and an arena of uncertainty, interpreted only through probability, and this gives rise to entanglement. In other words particles are entangled, joined up to each other in some way. But in this theory this entanglement is all compassing and accounts for 'the Arrow of Time' itself. It explains why our experience of time is of it moving forwards and not backwards – and very importantly does not rule out the possibility that time can move in both directions as suggested by recent experiments. We could theorize from Sandu Popescu's theory that direct-time possibly started when there was an event creating a dimension of direct-time – not a Big Bang but a place to evolve through movement and change.

Everything seeks an equilibrium with its surroundings. Equilibrium is all about pull and push and standing still once you are in place. However the complexity of the universe is such that nothing stands entirely still as it would cease to exist, which makes all movement relative and ephemeral.

The argument is that the impulse towards an equilibrium starts with the sub-atomic world and then moves on through the total environment. In the case of the coffee it firstly meets up with all the other particles in the room. As information is exchanged the inner complexities of the particle world lean towards an equilibrium with the total environment. It is a coffee, temperature, room, outside world process involving the total environment. This exchange of information causes the coffee to cool down but the rest of the room may still be continuing with the process of equilibrium. This they argue accounts for what we call time and its movement into the future.

The theory has many other implications and it is a very interesting article. For example, the process of reaching an equilibrium means time does not go backwards, although they add that this is not impossible but highly unlikely. This may relate to the almost incalculable number of possibilities. *They also point out that: "The universe as a whole is in a pure state," Lloyd said. "But individual pieces of it, because they are entangled with the rest of the universe, are in mixtures."* This is an idea which leaves room for different progressions as it implies that the world we experience is underpinned by the entangled and ever changing sub-strata always reaching for an equilibrium with the world out there. In short there are no straight lines to time and nothing is absolutely certain. That may well be the only way that chance, probability and coincidence can be part of reality.

Their theory is very relevant to my own perspective and it seems sensible to assume that the impulse to achieve an equilibrium relates to the peculiar nature of time. That is where the process starts and why the process is maintained. The reaching of stasis is a continuous ongoing process and the defining moment when stasis is achieved is a specific and definitive moment where time coheres. It a little like taking a picture. In fact you can take a picture of a picture of another picture, ad infinitum, and these are moments of stasis. I have just paused my video of match of the day. On the screen is an image of three people but one of them, Big Sam Allardyce, is from a previous year.

This brings me to another way of looking at time. This starts from the idea, as the evidence suggests that what we call space is a teeming seething and constantly changing mass that is forever sorting itself out and organising itself as it interacts and changes places. This constant changing can only be from one state to another. It is about movement as things jostle for position from the sub atomic to the particle world up. But we can to this our knowledge that some aspects of reality like light and other massless particles are able to ignore the constraints of the 'direct-time we, and matter in general, experience.

We can also add to this confusion what Feynman and others call retarded and future time. This is time running differently. We do not have retarded time or future time in our experience of reality but it is almost an established fact in the sub particle world of quantum mechanics. Entanglement, tunnelling, are all consequence of the peculiar nature of time. It appears to behave differently depending on which aspect of reality it is operating in. This leaves us with a mystery at the interface between two different notions of time which suggests we need to adopt the incredible idea that we need two notions of time. If they are correct then they have to be related and interdependent in some way so that one makes the other possible. Sandu Popescu and his team have provided us with a theory of time which may be pointing us in the right direction.

The area of reality where time can go backward or forward I want to refer to as core-time and that makes core-time the infrastructure of the universe. It is fundamentally at the inner reaches of the sub-particle world. Time as we know it, and as theorised as the search for an equilibrium, is the flow of the taken-for-granted world of us and our direct experience. It is direct and relates to change. It is built on the notion that space-time and movement is one and the same thing coexisting with the vacuum of 'core-time'. Nobody understands the vacuum and its workings but it is clearly an entity according to what we are learning from the world of physics.

Time is real but is a consequence of the mysterious workings and interrelationship it has with an underlying reality at its core. We need to think of core-time and direct-time as two different sides of the same coin. Science indicates this when it talks about uncertainty in the so called vacuum. Even allowing for probability it is unpredictable. Quantum theory see the vacuum as always being able to surprise us as many times as it likes and this has been well established through experimentation. It happens because, as the complexity of the theory dictates, and John Gribbin explains: "Quantum theory says that there is actually an entanglement between real photons and the photons of the vacuum itself." (6). In other words an entanglement between core-time and direct-time. It is a theory worth pursing if only that it opens up the issue of time itself.

Whatever core-time is it may continue to remain a mystery in our lifetime and beyond. We may suppose we dealing with an infinity, and why not as there is surely an infinity of something and mathematics keeps bumping into it. However, although I have referred to core-time as being infinite I must empathize it is not an infinity of nothing. It is an infinity of space, time and movement. Something infinite cannot cease to be but there is no reason why it cannot change into something else and why that change is a moment of time, mistakenly seen as packet or unit of time. It is transitory in an infinity of time. Remember, everything changes and nothing stays the same.

Core-time is programmed to interact with direct-time as we see in photosynthesis when core-time reaches out to direct-time and they behave as one. The relationship between core and direct-time is about the restoration of the equilibrium, the replenishment of time, or what David Deutsch would call the fabric of time, in this case, direct-time.

If there is a form of entanglement between these two aspects of time then we have a core-time which feeds into and manages direct-time by always reaching for an equilibrium. That makes the time we know a produce, an event or moment, something that happens in what we call real-time.

Seeing time as an entity is not a totally new idea. David Deutch, although viewing time differently, sees time as 'moments'. David is an advocate of the 'multiverse' and states that: *"We exist in multiple versions, in universes called 'moments'. Each version of us is not directly aware of the others, but has evidence of their existence because physical laws link the contents of different universes."* (7) This appeals to me as it is an idea that would not limit core-time to an interaction with one dimension and so the question is one of choice, one, two, ten or an infinity of dimensions?

We need to think about what the relationship might be between core-time and direct-time. They are two halves of reality, two different aspects of time. One, the sub particle world, gives rise to the other. In the first core-time, things behave differently and can move through time in any direction and in addition things can happen spontaneously, be in two places at once and tunnel across barriers as evidenced through experiments. In direct time things are different as, although being instigated through core time, there is stability and order. Logically we must assume a transition from one situation to another and ironically we see it as disorder to order when in fact it is order to disorder.

This idea of 'time' may sound strange but I would point it is no stranger than the information now coming at us almost daily from world-wide research and experimentation in physics. I remind you that the rules of tunnelling or spontaneity are about events in the particle world where there appears to be a different set of laws. This raises a question about the interface of two dimensions of time. We have to ask is how does a particle know when to behave spontaneously, or tunnel through barriers? Assuming it does not have a brain it can only take place when the particle knows it is in some way connected and has been placed in a position or situation relative to its other parts. Direct-time gives us a clue to that if we assume direct-time can be channelled in some way or given a direction. In other words if the connection is determined by an impulse to create order, in this case an equilibrium then what we have is the potential for an interwoven and continuous relationship.

It may well be a wild theory but I like the idea as it seems plausible that there is possibly a constant exchange between core-time and direct-time. The greater the movement needed for equilibrium the greater the role of core time. If you disturb the equilibrium by boiling water, making coffee or sunbathing you increase the rate of equilibrium and in so doing you speed up time. The process has opened up a wound, disturbed the equilibrium, and the repairs start pouring in. It is not like putting your hand in the river, it is more like putting the river into your hand. The equilibrium is at its most unstable, most transitory and more vigorous when things are heated up. In other words, when the equilibrium is disturbed, everything is moving about much faster and that means time is on the move. It never stops, is always moving, always being formed particle by particle, atom by atom, molecule by molecule from one moment to the other. Direct-time is relative to core time, which is not time standing still but the rate time is spewing out into the universe. Where a relative state of equilibrium is maintained the connection between core time and direct-time is at its slowest, the moment of stasis. Imagine time is standing still and so is movement. Now reverse the logic and have core time always on the move and direct time always seeking to stand still.

It is only as the force that bring about the search for an equilibrium takes hold that the world we live in begins to cohere that time becomes manageable, placed under control and given direction.

Our best candidate for this process is a force we are all familiar with but have no evidence for apart from its obvious effect on the world we live in, 'gravity'. However, we do have to enlarge its brief and recognise it is not just a force which brings things together but one that also functions to keep things apart by maintaining a state of equilibrium. And why not we have already conceded it had a different role in the so called Big Bang. This may sound like a piece of science fiction but it is actually a theory which relates to other work. John Gribbin mentions a theory of gravity based on the work of Shu-Yuan Chu a Californian professor of physics. Instead of 'fields' he suggest that:

"particles interact with each other in time symmetric way by exchanging advanced and retarded messages in a continues feedback and what we are used to thinking of as a continues field, such as gravity is built up by averaging over all the interactions involving little pieces of matter". (8)

Time, like the photon and other particles, is massless. This is not a problem in the particle world as a number of particles, like the photon, have no mass. What science calls mass in these cases is really just a reference to their force in the world of sub atomic particles. We know a force exists because the photons speed and direction can be changed and the energy of the photon can be transferred upon interaction with other particles. Time is similar as it has no mass, cannot sit still and has an unknown role to play in the vacuum. But in the world of direct-time it acts like a gravitational lubricant to maintain an equilibrium between elements of a particle world which would otherwise obliterate each other.

The implications are that Gravity is a form of energy which arranges matter, arranges the particle world by moulding it into an equilibrium that is, balancing out the forces of what science refers to as the vacuum of space. The particle world starts at the sub atomic core level where time operates under different rules we do not understand.

What we appear to have is an interrelationship between at least two aspects of a total reality, direct-time and core-time. It is not enough to assume that when we look it effects what happens, there is more to it than that and it is in the mysterious relationship we will find the answer. Hopefully we are looking in the right direction and are ready to think outside the box. It is essential to start by recognizing that things are connected in some way we do not understand. We need to entertain the idea it may have little or nothing to do with finding new particles. There is another type of energy which emanates from core-time and what we are talking about here is gravity.

CHAPTER TWELVE

Urban philosophy

The first and most obvious dimension, apart from our own, is the one where our history goes. We all have a history and it is not sensible to believe it exists only in our heads. It has to go somewhere and is possibly shared. History, information, cannot just disappear, ask any scientist or philosopher and they will tell you it's a fundamental law.

That means other times, or other dimensions of time are not only possible but essential. In that way everything can have its time and things can exist alongside each other in time, as everything does – including us in either a parallel universe or the branching world, we are about to discuss, of Everett and Wallace. Also, this idea of time as an accord, or movement towards equilibrium, does not undermine the idea, as Einstein pointed out, that time can be shaped and distorted, slowed down and speeded up, and now, as the latest research confirms, it can go backwards too.

The theory proposed here also implies that direct-time does not flow in a straight line forward or backwards as it is a process seeking an equilibrium. The time we take for granted is of this universe and as such it is always becoming and cannot stop. That makes our universe like a permanent simmering turbulent mass of an almost infinity of events and 'now' partly a construct of the brain and also a real but transitory event. We think and that is what we call 'now'. It exists in our heads. But there is also a 'now' out there which is forever in a state of flux. It is a world which would make no sense without us being able to create our own 'now'. It is when we are thinking and processing that we make 'now' a stationary and fixed moment – a permanent dimension outside which each event moves into the past apart from the 'now' we struggle to hold on to. It is we who grapple with now. It is we who construct it, hold now in our imaginations and release it into the past. We cannot see the future although we can guess at it, we cannot see the past except through our own tinted glasses. All we can see is 'now' and that is only possible through our imaginations - the working of our brains. But all the time the 'now' out there rolls on regardless.

If that is the case then there are real events taking place out there in reality but we select from that world as that is the only way to make sense of it. We create now but it is also a part of reality as a movement towards an equilibrium and, while not being completely observable, unless science is able to tell us differently, operates at an unobservable pace to all around it. So we see a solid reality and real events extrapolated from a moving world chasing a relative equilibrium.

This is a new concept of time that does not depend on a mathematical construct. It makes sense and more than that it makes room for the idea that there are other dimensions, which we will discuss later.

The theory makes time appear to be elastic and flexible and operating differently depending on whether it is responding to the reality of direct-time and its demands for order, or the demands of another aspect of reality in which its function is very different. Things do not just pop into time they are produced through a process best explained through quantum theory which we do not yet understand. They enter direct time and behave in a certain way. They cannot be timeless as all things change either by decay, collision or by changing into something else.

This is also a theory of time which would make room for the possibility that coincidence is an indicator of other dimensions and that in some way there is some form of interference or interconnectedness. The idea that there are other dimensions we do not know about is treated very seriously by many big name theorists. Apart from the ones I already have mentioned there are other scientist from John Wheeler and Richard Feynman to John Gribbin and Micheo Kaku who are convinced, and write convincingly, of other dimensions'.

Matt Strassler writes the following: ...*one of the ,most fascinating and non-obvious properties that the world might exhibit is that it may have additional ("extra") dimensions of space that you and I are unable to perceive, either directly through our senses, or indirectly through the many machines that we humans have built up to the year 2011. This possibility has been considered for at least 90 years, in various forms, and it is alive and well for physicist working at the large Hadron Collider, and beyond. (1)*

Before elaborating on my own ideas I must point out that they have been influenced by a number of people and I want to mention them in this final chapter. The first is David Deutsch and he takes on the challenge of extra dimensions in his book The Fabric of Reality. He points out that having another dimension is a way of explaining what is happening in the double slit experiment. He suggests it is not detectable by any other means, at this stage, than by the use of our imagination. I agree and that's is why I am allowing the notion of coincidence to steer a course for my ideas. He points out that all the experiments with light show that there is interference and this interference can only come from within the light itself. He accepts the argument that there is no such thing as empty space by asking the following:

'How does an object get from one place to the other if there is not a range of intermediate places for it to be on the way' (2)

He argues that experiments indicate that it is the behaviour of light and not the observer which accounts for the differences we observe in the two slit experiment.

In order to elaborate he hypothesizes what he calls shadow photons and tangible particles. The former belonging to a parallel universe and the latter being of this universe. Each photon is tangible in its own universe but not in the other. He also adds that because of the nature of interference there are far more shadow photons than tangible photons. It is not, as currently argued, that when we look we change things. What we are seeing is evidence of another reality or dimension of which there may be many.

Deutsch believes we are caught in an historical trap of looking through a logic we cannot let go of. In other words we have to face up to the reality of what we are seeing through our experiments with light – it bends and it interferes with itself and the only feasible explanation we can come to is that it is another undetectable dimension we are being confronted with. In short we have to accept that reality is a much greater mystery than we can imagine and we still have a long way to go before we can say we understand – if at all that will ever be a possibility. He goes on to state on page 45 of his book, *"the objects and events that we and our instruments can directly observe are the merest tip of the iceberg". (3)*

There is a site on Utube where you can visit another prominent scientist Micheo Kaku who shows us another way of seeing reality. He is a physicist of New York Institute for Advanced Studies and an highly respected figure in the world of science and points out that everything can be seen as a wave, me you and the tree in the garden which can be identified through the electromagnetic spectrum. That is, different wave lengths, depending on their frequency, being responsible for anything from radio to gamma rays. It suggests that the particle world is basically a receptor of these waves responding to incoming wavelengths and absorbing or emitting particles or other wavelengths. (4)

In our living room, he argues, we may be receiving a local radio signal but it is only one of a multitude of radio signal on the planet. We only hear the one in particular as we have de-cohered from the others. It is possibly the same for us, he argues, the possibility that we have also de-cohered with other dimensions. I remind you that Decoherence occurs when a wave collapse into a particle or one emerges from empty fog-like cloud of the atom.

His argument implies we also de-cohere from moments or events too. Life is a series of pictures and every second, just like a flash you see on TV, second by second we process these events. It is only through the brain and by intense concentration that we hold up the moment, stay in one or the other dimension for more than a moment and de-cohere from others. We can go many different ways as there are always dozens of routes but all are similar in content just like TV. It is an argument not too dissimilar to the theory of time I have outlined above, apart from seeing time as a described by Popescu and his team as a real event and not an illusion.

My advice to the reader is to keep an open mind, the governments of the world do and they invest heavily in innovation through swathes of people coming up with ideas even more outrageous and more extreme than mine. The reader should also make greater use of the media, especially search engines and documentaries – they really do offer an insight into modern day research. The CERN atom smasher has a regular output of up to date information available to all. (5)

I started off this journey by talking about coincidence and hoping to find some connection with events in the world of science. I must make it clear that I have no scientific answers about coincidence, only my intuition based on my readings. Clearly, my interest has shaped the outcomes of this journey but having read what is known and not known and a large number of strange, but respected, theories I have no fear about proposing another one of my own. But firstly I want to suggest that the most plausible is the multi-world theory of Everett and Wallace and propose that we call coincidence is often due to the inter-relationship of different dimensions. It is not a totally new idea as Wallace too sees decoherence and entanglement as the natural order of things and as the interface between all the different possibilities, the to-ing and fro-ing, the interconnectedness of the universes.

The multi-world, or Multiverse is a serious and respected theory very much linked to Hugh Everett the founder of the many-worlds interpretation. He would argue, and did, that we cannot ever know how it all began. Hugh Everett was at one time assistant to Neils Bohr, rubbed shoulders with Einstein and many others. Some, including Max Tegmark of CERN, see his contribution as important as Einstiens. And David Deustch was impressed enough to say he was 'before his time'. Everett argued that we are living in a multiverse and each one of these universes had a copy in it of each of us.

For Everett and others it resolves the problem of the two slit experiment and the idea that when we look we affect the outcome. In the sub atomic world of quantum mechanics everything exists differently to us. Things are neither one thing nor the other, just a potential to be. As such a particle can be everywhere at once, in this location or that location or as many locations as it likes. Everett presents this

197

foggy nebulous subterranean world as the clay of reality, the paint for the canvas, the soil of the garden so that when we intervene in it we create our reality. (6)

Everett did not like the Copenhagen idea of a superposition of states. Something was wrong with it and so he set about trying to see things from another angle. Having one result and not then other did not make sense. The answer, he thought possible lay in what we call measurement. Measurement is an act, something we do. It does not exist like an object and it is more an assumption that there is a relationship between objects, but what on earth is a relationship between objects? It is clear they do relate but science has never understood how and with quantum theory an understanding is now left to notions of force carriers and fields.

Probability means that if we measure everything going on we may get a rough idea of what may happen. In the sub atomic world everything is strange and incomprehensible and when we try to figure things out we can only start with the probable. How can anything exist with such un-certainty? When we measure or look we get a definite result but what happens to all the other possibilities? Do they now disappear from the scene, no longer exist. We say they collapse but that's because we don't know what happens to them. In fact as entities nothing should just be able to disappear. This leads to the question where did they go? The only theories we have are about things that behave themselves and do not disappear. These theories do not work at this ghostly area of the sub-atomic world where we see atoms like clouds with strange goings on in the middle. We need a new theory and we have to believe we are going to get one otherwise we are indeed no different than the Goldfish.

That is one reason why Everett theory is so popular but it also, according to those good at maths, makes sense. Everett saw the observer as part and parcel of the same world, at both a micro and macroscopic level. It was not just the particles of a sub-atomic world which are in a superposition of states but also you me and the moon. In other words nothing disappears, everything has a place and that place is in a multiverse. This universe grows like a tree and every time there is an alternative, a choice taking place, an intervention by an observer, a new branch is formed in the multiverse and heads of in a new direction. It is explained by the quantum idea of decoherence, the process whereby the wave collapses and the reality we see and touch takes place. It is a revolutionary idea but one that is very seductive and has added further to the debate. In a nutshell it is saying that what you see is seen differently by everybody else in the universe as every possible alternative view occurs. Incidentally, that also explains what is going on in the two slit experiment – both events occurred but you only established one of them in your universe.

David Wallace Professor of physics at University of Oxford has written a book, the Emergent Universe which further develops Everett's theory of the multi-world. (7) He too points out that Quantum mechanics is telling us that there is more than one outcome to an event and is just like any other theory about the world out there. All we have to do is accept what it is telling is us. A superposition of two or more possible states means a particle, a fundamental bit of matter, can be in two places at once and have different spin. Amazingly, it is only when we try to pin matter down by measurement or staring at it that we make it behave itself. That at the moment appears to be the process by which the world is assembled.

Being in two places at once is now an established fact and scientist are hoping to use such knowledge to build a super computer which will make use of extra mathematical possibilities. A piece of recent research in Australia, by a team led by Andrew Dzurak and published today suggests it is really going to happen. (8)

To make sense of what is going on the best theory they have is based on what they call the 'wave function'. A wave is something that goes up and down and rolls along and they all have a frequency. That is its wave function. Physicists use this wave function as a means of figuring out the 'probability' of where the particle is. But there is a problem for the theory as pointed out by the famous mathematician H Poincarre:

"The very name calculus of probabilities is a paradox. Probability opposed to certainty is what we do not know, and how can we calculate what we do not know? (9)

There is another way of putting this. Imagine a pool and in that pool is an animal which will briefly come up for air. The certainty is that it will come up in that confined and limited space. Once you break the area of the pool down into sections you think it may appear in you have introduced uncertainty. Before you did that you had certainty - there was going to be an appearance. Now imagine you do not know which pool, if any, it will appear in. Now nature has introduced uncertainty.

Clearly we need to give some thought to what is actually meant when we measure probability. It is a contentious issue in the world of science and is taken up by Wallace. He points out the contradiction in research in the rest of science. You are not measuring something that is there, rather you are trying to measure something that may or may not be there. It is not and cannot be the same as pinning something down with certainty.

He sees probability as a problem for all the disciplines and points out that it is a mystery. When using probability any outcome is always made in some sort of context and is based on an algorithm or set of rules. The fact that the outcome is probable rather than actual suggests that something breaks the symmetry but we just assume chance. What on earth is chance? Why one outcome and not the other. Is probability the reason we have chance or is chance the reason we have probability? In the subatomic world some atoms can decay at any second and that is the most probabilistic statement we can make about them. There is definitely room for coincidence as a key player here but I am not ready to pursue that yet.

Nobody understands probability, this is true in all branches of science. Mathematicians have sleepless nights over it as it reminds them of our limitations. Using probability to make sense of the world is no different than guessing. Wallace uses the example of a dice. Only one outcome can occur with each throw and we use probability to explain this happening but he points out that while only one outcome actually occurs the other possibilities also occur. This is based on the notion that probability, apart from not being a real thing, is a flawed concept in that the chances are not objectively valid and the choices you get are related to something in the vagueness of the subatomic world.

Wallace, Everett and others believe that the outcome is related to the interconnectedness of the multi-world. Another way of viewing this problem of probability is to imagine somebody with no knowledge of human life seeing only ripples or dots on a screen when looking at what we know as a street. They may be able to predict how many people will appear in the street for any given period based on the collation of previous data. You can take measurements yourself over a period of time and build an algorithm or set of rules with all your information. However, your predictions will say nothing about the people in the community, where the come from, go to, interrelate, die, commit suicide and have families, there relationships to each other. That is also the situation for the Copenhagen interpretation. It says nothing about the underlying reality which in this case is labelled by Everett and Wallace as the multi-world.

The many-worlds theory is telling us that there is another direction we can look if we want to discover the underlying laws by which the universe operates. It's an amazing and stimulating theory! And I am very sympathetic to the idea. It involves a process called 'branching' in which two things emerge out of one, and out of this emerges another four and then on and on in infinite multiplication as things move further and further apart. Branching occurs every time a choice is made. It is a bit like an analogy with a chess board and doubling the amount of seeds on each square. It is a neat idea as it mirrors the development of the individual, the family and the notion that there is a universe for each of us.

There is no reason, as String Theory suggests, why we cannot assume there are a number of times or dimensions. There may even be an infinity of times. However, whether an infinity of time or an infinity of a number of dimensions of time they cannot be infinities or dimensions of nothing. For information and history never to be lost time would have to be infinite. Unlimited possibilities can only exist in some form of infinity or in an infinity of other times or dimensions. For core-time to be a part of that means what we call core-time is possibly a potential conduit underpinning and maintaining the direct-time of this and many other dimensions.

If that is the case then movement into a future time would be an impossibility as the future in the multiverse consists of infinite possibilities. That makes core-time a potential route across times and that means there is no going directly back in time. There is no straight line into the future or back to the past. That would eliminate the problem of changing the future and negating your own birth as all that would be created would be a new path into the future. This does not negate the possibility of movement through time as any moment in time is a sideways movement enabling a snakelike path through time.

The idea of infinity of events would answer a lot of fundamental questions I will not pursue here. That is not a good enough reason for believing it correct even though it feels intuitively a direction we should certainly look in to develop a greater understanding. Sometime we must follow our intuition and listen to our subconscious and so I too believe that we are one universe among many but not necessarily an infinity of dimensions. The theory is impossible to prove. Coincidence, or the probability of this or that happening may be the best chance we have of unravelling the mystery as it is only possible to give substance to the theory through the existence of one or other dimensions.

Once you start thinking about different dimensions of time you are left thinking what a narrow egotistical idea it is to think there is only one dimension of real time. If we play about with the idea of core and direct-time then a dimension of direct-time emanating from a single dimension of core-time seems a bit weak. We are talking about an area of time we know very little about except that it behaves totally different to anything else we have ever come across. It is a part of our reality we are only able to guess at. It is a theory so filled with potential for the future that it would be sensible to ask, why one direct time and not two or more times?

I did say I was not afraid to propose my own theory and the more I go on the more adventurous I get. So far all the ideas discussed so far have been based on the theories of others but I would now like to propose one that apart from the notion of 'tunnels in time' is a work of pure fiction. Perhaps what I have called core-time is exchange time. It is like the centre of spokes in a wheel or a station waiting room for things passing between dimensions. Core-time is not something we can call an entity on its own it is more like a process between the different dimensions it feeds. Events in the sub-atomic world happen at this crossroad of time. Core-time runs in all directions within its own dimension where it is incomprehensible but possible exists only because of the way it is fed into dimensions. That would suggest reality and core-time are two halves of a complete reality in a multidimensional reality.

If there are other dimensions and there is some form of communication between them then it would seem logical to suppose that whatever other dimensions are involved some will be more important than others if only because of the way they are related through time. Such dimensions would be most important as they are more likely to provide indications or footprints in the form of coincidences which may actually lead us to some evidence of the theory.

Whether there is or there is not a multiverse I am convinced that the theory of core-time and direct-time is an important step in the right direction. In the two slit experiment what we see happening is both times at the same time until a choice is made. At that moment we make equilibrium a real possibility. However, it is not just our observation which brings this about, it is also any potential interference which can trigger the process as the information coming from the biological world of science in photosynthesis indicates. A dual notion of time makes sense of the scientific evidence now available and points in a new direction.

This all suggests, as I pointed out earlier, that what we call the Big Bang was more a form of entryism, a projection into a dimension by the creation of time. We should be looking for what created time.

I remind the reader that this is a theory based on the idea that it is impossible to imagine two spontaneous events, millions of miles apart, in a single dimension of time, impossible unless they are running alongside each other (as everything is moving) but are in separate and interconnected dimensions and linked to the same time. How else could a spontaneous event occur between two identical entities? Two tops become bottoms, lefts become rights etc. In other words for spontaneity to take place time would have to bridge both dimensions.

The actual connection between core-time and the dimensions it feeds is a mystery. But whatever it is we must make room for coincidence or the idea of a quantum option to start to make sense of it. Coincidence is possibly a form of de-coherence, an entanglement of its own in the process of bringing about events across the universe. It is capable of occurring at any time and in some cases more frequently than others although we would appear to have no control over these events.

My last word on this involves a touch of irony as the latest coincidence of significance comes from CERN itself. On the 17th of December 2015 they issued a statement about the possible discovery of a new particle. (10) It was a finding that caused quite a stir as it had also been spotted not only in the Atlas experiment at CERN but also at a completely separate experiment called CMS (Compact Muon Solenoid) one of two particle physics detectors at CERN. It is built in an underground cavern at Cessy in France and was also involved in the discovery of the Higgs-bosun. Part of its brief is looking for other dimensions and the make-up of dark matter.

However, by summer 2016 it was announced as a statistical fluke, or as Matt Strassler puts it, A Flash in the Pan. Officially it was explained away as a 'coincidence'.

Other dimensions is an incredible idea and can be seen as a wonderful moment of human comprehension. The history of the human race and its development is built on such moments. We appear to have the ability through acts of imagination and comprehension to challenge the unknown through outrageous thought. We can do that because we have consciousness, the greatest mystery of all!

We can be conscious of all things, but all things cannot happen at the same time in a single dimension. It is only by straddling dimensions that spontaneity can occur and that is the door we must knock on to have any future understanding of reality.

jaybee

APPENDIX - BIBLIOGRAPHY

CHAPTER 1

1. Arthur Koestler. The Roots of Coincidence. (1972), US, Random House hardcover:

CHAPTER 2

1. Coincidence or Destiny. Stories of Synchronicity That Illuminate Our Lives by Phil Cousineau. Conari Press. Paperback (2002) (pages 85 – 86)
2. The Power of Coincidence. The Mysterious Role of Synchronicity in Shaping Our Lives by Frank Joseph. Paperback. Mar 2009 pages 139/140.

3. Jung on Synchronicity and the Paranormal. Carl Gustav Jung. Roderick Main. Psychology Press, 1997. You can read about Jung under 'Examples' on the following website:
http://en.wikipedia.org/wiki/Synchronicity

4. The Lost Paradigm of Coincidence. Article by John Townley and Robert Schmidt. A summary of The Inheritance of Acquired Characteristics. Paperback – October 2012. By Paul Kammerer.

CHAPTER 3

1. Live Science. Feb 2014. Exotic Particles, Tiny Extra Dimensions May Await Discovery. By Katia Moscovitch.
http://news.yahoo.com/exotic-particles-tiny-extra-dimensions-may-await-discovery-163233696.html

2. The Structure of Scientific Revolutions. University of Chicago Press by Thomas S Kuhn 1996

3. The Logic of Scientific Discovery (Routledge Classics) Paperback Feb 2002 by Carl Popper Author.

4. Richard Feynmans. Penguin Books. The Strange Theory of Light and Matter: page 119

5. YouTube. How Lasers Work (In Theory). Minutephysics. Published Dec 2011

https://www.youtube.com/watch?v=y3SBSb
sdiYg

Or How Lasers Work. Animation with Albert
Einstein. Published 2014 Thomas Schwenke.
https://www.youtube.com/watch?v=1LmcUaWu
Yao

6. Life on the Edge. The coming of Age of
Quantum Biology by Jim Al-Khalili and
Johnjoe McFadden Bantam Press 2014 (pages
10/11)

CHAPTER 4

1. John Gribbins Schrodingers Kittens and the
Search for Reality. Weidenfield and
Nicholson. Orion Books Ltd (1995) pages 1-
9.
Or an Excellent site:
*https://www.youtube.com/watch?v=DfPeprQ7
oGc*

2. A Brief History of Time: From |Big Bang to
Black Holes. By Stephen Hawkins. Paper
back 18 Aug. 2011. Random House
Publishing Group. (Pages 74-75)

3. John Wheelers thought experiment:
http://www.unt.edu/rss/class/rich/misc/JohnW
heeler.html

4. Delayed choice proved:
http://www.ibtimes.co.uk/quantum-

weirdness-proved-again-measurement-changing-atoms-past-1504172.
Delayed choice. Nature Physics. Experiment Shows Future Events Affect the Past. By Paul Radcliffe July 9th 2015. Also see the following:
http://scienceandnonduality.com/experiment-shows-future-events-affect-the-past/#sthash.barGU8yG.dpuf

5. Type 'Alain Aspect Experiment' into Google for a choice of videos. One about the actual experiment done by Alain Aspect (1981) and another an excellent version of the debate between Einstein and Neil's Bohr.

6. Entanglement - Spooky https://www.youtube.com/watch?v=BFvJOZ5 1tmcction at a Distance. A public release on the net about research at the University of Vienna titled 'Photons Run out of Loopholes' Eureka Alert. 15 April 2013. http://www.eurekalert.org/pub_releases/2013-04/uov-pro041513.php

Chapter 5

1. Video on Utube, The Standard Model: https://www.youtube.com/watch?v=V0KjXsG RvoA

2. The basics of particle physics. http://www.learning-

mind.com/understanding-the-basics-of-particle-physics/
3. The Higgs Field analogy: http://cds.cern.ch/record/1458922
4. Beyond Higgs: http://news.yahoo.com/beyond-higgs-5-elusive-particles-may-lurk-universe-121157049.html
5. Listen to Professor Peter Higgs Radio 4 broadcast here: http://www.bbc.co.uk/mediacentre/latestnews/2014/radio4-life-scientific-peter-higgs.html
6. Alan Alada. Scientist need to cut out the jargon: http://phys.org/news/2013-05-alan-alda-scientists-jargon.html
7. Professor Michio Kaku website: http://mkaku.org/
8. Conversations about Science with Theoretical Physicist Matt Strassler. Of Particular Significance. Website: Mattstrassler.com
9. Takaai Kajita and Arthur McDonald get the Nobel prize in physics.

See what it's all about here: http://www.onenewspage.com/n/Technology/7559xgi62/Nobel-Prize-in-Physics-honors-discovery-of-neutrino.htm

CHAPTER 6

1. Wired.com. The Future of Physics Explained.

http://www.wired.com/2012/07supersymmetry
-explained/

2. Brian Greene String Theory:
http://www.ted.com/speakers/brian_greene

3. Lawrence Krauss interviews Brian Greene
about string theory. It's on Utube
ScienceNET 5th June 290014
https://www.youtube.com/watch?v=3kn5XK
KZEFU

4. Micheo Kaku on String Theory.
https://www.youtube.com/watch?v=6wGPR1
4ZAss

5. Wn.com. Fine Structure Constant –Sixty
Symbols. April 2009. Excellent video
http://wn.com/fine_structure_constant

6. Professor Jordan Nash:
http://www.bbc.co.uk/news/science-
environment-14680570

7. BRANES. The Mother of all Super strings.
http://mkaku.org/home/articles/m-theory-the-
mother-of-all-superstrings/

8. The TV series can be seen at this website:
http://www.sciencechannel.com/tv-
shows/through-the-wormhole/

9. David Deustch in his book The Fabric of
Reality, The Penguin Press 1997 (Page 328)

10. Conversations About Science with
Theoretical Physicist Matt Strassler. Of
Particular Significance. Website:
Mattstrassler.com

http://profmattstrassler.com/2014/03/07/what-if-the-large-hadron-collider-finds-nothing-else/

CHAPTER 7

1. The Light Behind Consciousness. John Archibald Wheeler. Publisher non-duality Press September 2008
2. The Holographic Brain with Karl Pibram, Ph.d. A transcript from the series Thinking Aloud. You can read this at the following web site: http://intuition.org/txt/pribram.htm
3. David Bohm proposes another model, The Lens defined model and you can read more about it at this site: https://thestillmind.wordpress.com/tag/david-bohm/
4. Sir Roger Penrose and Stuart Hameroff argue that we cannot dismiss the possible effects on conscious, in particular the collapse of the wave (https://www.youtube.com/watch?v=OfdsGpTHIkw)
5. Godel's Wikipedia. The classical Liar Paradox. https://en.wikipedia.org/wiki/Liar_paradox
6. Anirban Bandyopadhyay: quantum vibrations inside microtubules: http://phys.org/news/2014-01-discovery-

quantum-vibrations-microtubules-corroborates.html

7. Life on the Edge. The coming of Age of Quantum Biology by Jim Al-Khalili and Johnjoe McFadden Bantam Press 2014 (Page 134)

8. : Elisabeth Rieper at the University of Signapore. http://www.popsci.com/science/article/2010-06/quantum-entanglement-may-hold-dna-together-new-study-says

9. 'Conectomes'. Professor Sebastian Seung, professor at Massachusetts University maps the connections in the brain http://www.nytimes.com/2015/01/11/magazine/sebastian-seungs-quest-to-map-the-human-brain.html?_r=0

10. The Spooky World of Quantum Biology: *http://realitysandwich.com/29035/spooky_world_quantum_biology/*

11. Electron paths: *http://phys.org/news/2013-06-uncovering-quantum-secret-photosynthesis.html*

12. Neorons and synapses – memory and the Brain -The human memory. F: http://www.human-memory.net/brain_neurons.html

13. Stanford Medicine. News Centre. By Bruce Goldman. Published June 11th in Science.

STANFORD:
http://med.stanford.edu/news/all-
news/2015/06/genetic-underpinnings-of-
functional-brain-networks.html

14. On the basis of the study Theoretical Physicist George Rajna has submitted a paper to Academia.edu titled The Genetic Basis of Brain Networks.
https://mail.google.com/mail/u/0/#inbox/14e1
1d0ebe47ad04

15. Morgan Freeman: 47 Utube videos of Through the Wormhole series.
http://www.bing.com/videos/search?q=m
organ+freeman+through+the+wormhole&qpv
t=morgan+freeman+through+the+wormhole&
FORM=VDRE

16. Ian Sample on Jeff Lichtman. The Guardian Newspaper.
(8th may 2012)

17. Johnjoe McFadden, Quantum Evolution, Published by HarperCollins Quantum Evolution: Life in the Multiverse Paperback – 4 Mar 2011.

18. Digital Market. 95% Of our thoughts and decisions occur within the subconscious: Priyanka Bhattacharya. August 2012.
http://www.digitalmarket.asia/95-percent-of-
our-thoughts-and-decisions-occur-within-the-
subconscious/

19. BBC. News Science and Environment. Do You See What I See? August 2011. Also shown on BBC Horizon 2011 – 2012 http://www.bbc.co.uk/news/science-environment-14421303

CHAPTER 8

1. Inflation for beginners. John Gribbin. http://aether.lbl.gov/www/science/inflation-beginners.html
2. Robert Brandeemberger. Big Bang, Deflated? Universe May Have Had No Beginning. By Tia Ghose, Staff Writer. Feb 2015. http://www.space.com/28681-theory-no-big-bang.html
3. The Planck Era http://csep10.phys.utk.edu/astr162/lect/cosmology/planck.html
4. 1929. Edward Hubble Discovers the Universe is Expanding. http://cosmology.carnegiescience.edu/timeline/1929
5. NEW SCIENTIST. How Dirac predicted antiumatter. May 2009 by Roger Highfield
6. Anti Matter. Cosmos and Culture. June 2011. A great mystery comes into focus. Anti matter trapped for,16 minutes.

7. March 2014. Astronomers discover echoes from expansion after Big Bang. By Irene Klotz and Sharon Begley. New York.
http://www.reuters.com/article/2014/03/17/us-science-bigbang-idUSBREA2G16F20140317

8. The Guardian. Ian Semple. Science Editor, Gravitational waves turn to dust after claims of flawed analysis.
http://www.theguardian.com/science/2014/jun/04/gravitational-wave-discovery-dust-big-bang-inflation

9. Picture of WMAP Cosmic Microwave Radiation
http://map.gsfc.nasa.gov/

10. There is also an excellent TV programme about the issues on Horizon BBC2. July 2015

11. Gravitational waves Professor John Barrow. Gresham College
http://www.gresham.ac.uk/lectures-and-events/gravitational-wave-astronomy

12. Article in HAARETZ. June 2015. A quadrillionth of a second (specifically, 10 -34 second, or a decimal point followed by 33 zeroes and a 1).
http://www.haaretz.com/life/science-medicine/1.580346

13. Venezianos article. Before the Big Bang.
New Scientist. June 3, 2000

CHAPTER 9

1. Leanard Suskind Morgan freeman
   THROUGH THE WORMHOLE
2. ARS TECHNICA Scientific Method. How
   a disagreement with Hawking suggested
   the universe is a hologram. John Timmer
   July 2011
   http://arstechnica.com/science/2011/07/ho
   w-an-argument-with-hawking-suggested-
   the-universe-is-a-hologram/

3. TV programme The Observable Universe,
   Horizon Sept 2012
4. NATURE. Jan 2014. Article by Zeeya
   Merali. Black holes may not have an
   'event horizon' after all.
   http://www.huffingtonpost.com/2014/01/24/st
   ephen-hawking-black-holes-event-
   horizons_n_4658220.html

5. An excellent video about Vera Rubin. The
   Dark Side of the Cosmos. Heather Couper
   presents a narrative history of astronomy.
   http://www.bbc.co.uk/programmes/b00c76
   bn
6. DANCING IN THE DARK –THE END

OF PHYSICS March the 17th St Patricks Day 2015
7. FACTS ABOUT UNIVERSE. NASA WMAP
http://map.gsfc.nasa.gov/universe/uni_matter.html
Also see: DARK ENERGY, DARK MATTER. NASA SCIENCE
http://science.nasa.gov/astrophysics/focus-areas/what-is-dark-energy/
8. Not sure about Dark Matter:
http://profmattstrassler.com/2015/04/13/dark-matter-how-could-the-large-hadron-collider-discover-it/
9. Virtual particles NEED A SITE
10. Dark Matter. Melvyn Bragg and his guests discuss dark matter the 'missing mass' of the universe. BBC Radio4 (12/03/15).
11. WIMPS: Is Dark Matter Real. Space.about.com. Basics of Astronomy and Space Exploration.
http://space.about.com/od/astronomybasics/a/Is-Dark-Matter-Rcal.htm

12. Origins of the Polar Jet Seen for the First Time. Sept 2012.
http://physicsworld.com/cws/article/news/2012/sep/27/origins-of-galactic-jet-seen-for-the-first-time

13. Colin Stuart. Mysterious 'dark flow' at the edge of the universe: Physicsworld.com. http://physicsworld.com/cws/article/news/2010/mar/15/mysterious-dark-flow-at-the-edge-of-the-universe

CHAPTER 10

1. Buckmaster fuller http://www.goodreads.com/quotes/13119-you-never-change-things-by-fighting-the-existing-reality-to

2. The Ipaper. The essential Daily Briefing. From The Independent 6/05/2015

3. Life on the Edge. The coming of Age of Quantum Biology by Jim Al-Khalili and Johnjoe McFadden Bantam Press 2014 (page 120)

4. Life on the Edge. The coming of Age of Quantum Biology by Jim Al-Khalili and Johnjoe McFadden Bantam Press 2014 (page 129).

5. Chaos Theory http://fractalfoundation.org/resources/what-is-chaos-theory/

6. Guardian Newspaper on the 4th of November 2015. Craig Hogan a theoretical astrophysicist at Fermi-lab The 4th of November 2015. For more information visit nautil.us

7. WISP: MIT Technological Review. New Kind of Quantum Tunneling Experiment Goes Live. Physicists may be able to prove the existence of Dark Matter by watching a blank wall. *http://www.technologyreview.com/view/413722/new-kind-of-quantum-tunneling-experiment-goes-live/*

CHAPTER 11

1. David Deustch in his book The Fabric of Reality, The Penguin Press 1997 page 47.
2. The Information Philosopher. Nicholas Gisin (1) *http://www.informationphilosopher.com/solutions/scientists/gisin/*
3. Logical Paradoxes http://www.logicalparadoxes.info/arrow/
4. John Gribbin Schrodingers Kittens and the Search for Reality. Weidenfield and Nicholson. Orion Books Ltd (1995) page 101. And Richard Feynmans. Penguin Books. The Strange Theory of Light and Matter: page 119. Also http://scienceandnonduality.com/experiment-shows-future-events-affect-the-past/#sthash.barGU8yG.dpuf
   Also read the very latest research on the following site: It was achieved in a recent experiment by an Australian team and

published in May 25 2015 in Nature Physics

5. New quantum theory could explain the flow of time. By Natalie Wolchover. Quanta Magazine. Popescu, Short etal. *Original story reprinted with permission from Simons Science News, an editorially independent division of* SimonsFoundation.org *whose mission is to enhance public understanding of science by covering research developments and trends in mathematics and the physical and life sciences.* http://www.whatsbestforum.com/showthread.php?14540-New-Quantum-Theory-Could-Explain-the-Flow-of-Time

6. John Gribbin Schrodingers Kittens and the Search for Reality.   Weidenfield and Nicholson. Orion Books Ltd (1995) Pages 121/122

7. David Deustch in his book The Fabric of Reality, The Penguin Press 1997 page 287.

8. Shu-yuan. John Gribbin Schrodingers Kittens and the Search for Reality. Weidenfield and Nicholson. Orion Books Ltd (1995) page 231

CHAPTER 12

1. EXTRA DIMENSIONS> http://profmattstrassler.com/articles-and-

posts/some-speculative-theoretical-ideas-for-the-lhc/extra-dimensions/

2. David Deustch in his book The Fabric of Reality, The Penguin Press 1997 page 36
3. David Deustch in his book The Fabric of Reality, The Penguin Press 1997 page 45
4. Micheo Kaku: Parts of Me Ooze in all Directions. Utube https://www.youtube.com/watch?v=gIieq0a130c
5. Up to date information from CERN. Accelerating Science.
   .   http://home.web.cern.ch/

6. Six Hugh Everett Biography. http://www.scientificamerican.com/article/hugh-everett-biography/
7. David Wallace, *The Emergent Multiverse: Quantum Theory according to the Everett Interpretation*, Oxford University Press, 2012.
   Reviewed by Peter J. Lewis, University of Miami
   http://ndpr.nd.edu/news/38878-the-emergent-multiverse-quantum-theory-according-to-the-everett-interpretation/
8. Andrew Dzurak. 6th October 2015. http://newsroom.unsw.edu.au/news/scienc

e-technology/australian-teams-set-new-records-silicon-quantum-computing

9. H. Poincare. Science and Hypothesis. Cosimo Classics, 2007, Chapter X1. More on Poincarre: www.informationphilosopher.com/solutions/scientists/poincare

10. Source: CBCNEWS –Technology and Science. Emily Chung CBC News Dec 17 2015. http://www.cbc.ca/news/technology/new-particle

'New particle' found at Large Hadron Collider wasn't for real - CBC